CAPTURED ARI CARS AND O' VEHICLES

In Wehrmacht Service in World War II

Werner Regenberg

translated from the German by David Johnston

A Humber Mark I armored scout car in service with a German unit in North Africa. The unit has equipped itself completely with English equipment. The Quad tractor in the right background is also marked with German crosses. (BA)

Schiffer Military/Aviation History
Atglen, PA

FOREWORD

This volume is dedicated to armored cars and other armored vehicles of various nations in German service during the Second World War.

Some of the photographs reproduced in this volume are lacking in clarity due the original photographic equipment used to take them or the photo source. They were used nevertheless on account of the rarity of the subject vehicles.

The author is grateful for any corrective or supplemental information in word or photograph.

Dr. Werner Regenberg

Left:
An abandoned or knocked-out Panhard armored car. These vehicles were recovered after the conclusion of fighting and if possible put back into service following a general overhaul by the manufacturer. (HH)

TITLE ILLUSTRATION:
Marmon-Herrington armored car (see Page 27)

PHOTO SOURCES

Bundesarchiv (BA)
Munin Verlag (MV)
Richard Eiermann (RE)
Reinhard Frank (RF)
Henry Hoppe (HH)

Friedrich Husemann (FH)
Randolf Kugler (RK)
Thomas Lederer (TL)
Stephan de Meyer (SM)
Karlheinz Münch (KM)
Palicki Archive (AP)

Jean-Louis Roba (JR)
Walter Spielberger (WS)
Prof. Dr. Manfred Savodny (MS)
Jürgen Wilhelm (JW)
Title Page: Heinz Rode

Translated from the German by David Johnston.

Copyright © 1996 by Schiffer Publishing, Ltd.

Printed in China.
ISBN: 0-7643-0180-2

This book was originally published under the title,
Waffen Arsenal-Beutepanzer unterm Balkenkreuz: Panzerspähwagen und gepanzerte Radfahrzeuge
by Podzun-Pallas Verlag

We are interested in hearing from authors with book ideas on related topics.

Published by Schiffer Publishing Ltd.
77 Lower Valley Road
Atglen, PA 19310
Phone: (610) 593-1777
FAX: (610) 593-2002
Please write for a free catalog.
This book may be purchased from the publisher.
Please include $2.95 postage.
Try your bookstore first.

INTRODUCTION

The usual roles of the armored car and the armored wheeled vehicle are reconnaissance, communications and escorting transport. Their military roles were described as follows in the German Army Manual D 612/1: "Because of their high speed, armored cars (variously also called road tanks, armored trucks, armored scout cars or armored motor vehicles) are used mainly for reconnaissance. Not only does their armament and armor enable them to reconnoiter in the face of opposition, but also to drive away enemy reconnaissance forces and exploit favorable opportunities for ambush. They shall also protect motorized units while on the move and in special cases themselves provide safe transportation for persons, equipment or intelligence information."

Outside Germany armored cars were usually classified as heavy or light.

Heavy armored cars were always equipped with an antitank weapon and machine-guns and most had radios. In general they weighed more than seven tons. In order to achieve a high degree of off-road mobility, in many cases they used special chassis or improved commercial multi-wheeled chassis. Some were equipped with rims for rail travel, which enabled them to be used as armored track motor cars. Light armored cars were mostly produced from commercial chassis, which were often only partially armored and then armed only with machine-guns. Often they were restricted to hard-surfaced roads.

In some cases the designation was reflected the type of armament ("*Autocanon*" or "*Auto-Mitrailleuse*") or the nature of the running gear ("street tank").

With motor vehicles in general use, the turn of the century saw the birth of the idea of using them for military purposes. The first experiments with armored trucks were carried out by the English in 1900 and the French in 1901. The vehicles in question were trucks fitted with makeshift armor and armed with machine-guns.

In Germany, in 1903 Paul Daimler began development of an armored car with four powered wheels, and a single example was built by the Austrian Daimler-Motoren-Gesellschaft in 1905. Only a few armored cars were built prior to the First World War, mainly by the French, but with the war they came into more common use, which stimulated a constant development of technology and tactics.

By the beginning of the Second World War all participating nations had armored wheeled vehicles produced at home or purchased abroad.

By the start of the Second World War the German Armed Forces had developed and introduced into service various types of light and heavy armored wheeled vehicles for its motorized units. These were the Kfz. 13 machine-gun car, the Kfz. 14 radio car, the SdKfz. 221, 222 and 223 light armored cars, the six-wheeled SdKfz. 231 and 232 heavy armored cars, the SdKfz. 263 heavy armored car, and the eight-wheeled SdKfz. 231 and 232 heavy armored cars.

In the course of the war this diversity of types was joined by the SdKfz. 260 and 261 small radio cars, the SdKfz. 233 eight-wheeled heavy armored car, the SdKfz. 234/1 to 234/4 heavy armored cars, and several armored automobiles and trucks.

More than 4,200 of the above-named types were built during the war years. However production of these armored wheeled vehicles never sufficed to fully equip all the units for which such vehicles were planned. Many units therefore had to make do with automobiles or even motorcycles in place of armored cars. Still-serviceable captured armored cars and armored wheeled vehicles were often added to the inventory straightaway, even in cases where the unit was not intended to have armored vehicles. Other units used captured vehicles

The state of this Russian BA-10 is typical of found enemy vehicles. The armored car was probably abandoned because of a flat tire. The crew or other passing troops removed the bow machine-gun and helped themselves to spare parts from the motor. The warning inscription "breech loaded" suggests that a live round was placed in the breech, which was then rendered unusable. (HH)

German troops inspect a captured Panhard armored car. The antenna on the left rear of the vehicle indicates that it is equipped with a radio. Not all French armored cars had radios as adequate numbers of sets were not available.

to fill gaps in their authorized inventories. The vehicles were used in their designed functions, as tractors for light towed loads, or as armored transport vehicles.

Planned formations of units in platoon to battalion strength took place with captured armored cars of English, French, Dutch, Italian and Russian origin.

Before and during the war the Heereswaffenamt (Army Ordnance Office) collected material on the weapons of other nations and produced so-called "Foreign Equipment Reference Sheets." In the process a system was developed in which the German name for the type of equipment came first, fol-

lowed by a reference number and a letter in brackets indicating the nation of origin. Numbers in the 200 range were used for armored wheeled vehicles, and the designation for the French *Automitrailleuse de Dévouverte Panhard 178*, for example, was:

Panzerspähwagen P 204 (f)

In this volume captured wheeled vehicles are dealt with by country of origin, with an effort being made to follow the progression of the war and thus the time of introduction; however this was not possible in every case.

A Polish wz. 34 scout car destroyed in combat. The armored car was armed with a 37-mm cannon which was housed in a ball mount.

4

POLAND

At the outbreak of war in September 1939 the Polish Army had approximately 90 armored cars, which were assigned to the cavalry brigades in units of eight vehicles.

Only eight of the *Samochód pancerny wz. 29* (Armored Car Type 1929) were on hand. This armored car had been developed at the end of the nineteen-twenties and approximately 20 examples were built. The vehicle had a crew of five (including a front and rear driver) and, weighing 4.8 tons, reached a speed of 35 kph. The armored car was armed with a French 37-mm Sa. 18 Puteaux and two 7.92-mm Hotchkiss machine-guns in the turret. One of the two machine-guns was installed in the roof of the eight-sided turret for anti-aircraft defense. A third Hotchkiss machine-gun was mounted in the superstructure facing toward the rear. The vehicle's armor was four to eleven millimeters thick. The eight armored cars of this type still on strength when war broke out were assigned to the Mazowiecki Cavalry Brigade as an armored scout squadron. The rather elderly vehicles were underpowered and had only a limited off-road capability. The wz. 29s, which were equipped with an Ursus motor and were also known by the name "Ursus Armored Car", were knocked out in combat or were destroyed by their own crews after becoming stuck in difficult terrain.

The second type of Polish armored car, the *Samochód pancerny wz. 34*, was developed from the French Citröen-Kégresse B-10 half-track vehicle. The Polish Army acquired a large number of these vehicles. Ninety were fitted with armored bodies and introduced into service as the *Samochód pancerny wz. 28*. As the vehicle's rubber tracks had only a limited service life, the vehicles were reequipped with a new rear axle with dual wheels instead of the tracked running gear and were redesignated wz. 34. The armored car had a crew of two and reached a speed of 55 kph. Armor thickness was eight millimeters. Armament consisted of either a 7.92-mm Hotchkiss machine-gun or a 37-mm SA. 18 Puteaux cannon in a revolving turret. About a third of the armored cars were armed with cannon, the rest with machine-guns. Many of the armored cars were lost because of technical defects and lack of fuel, others became stuck on account of their limited off-road capabilities and had to be abandoned. So far there is no evidence to prove that Polish armored cars were used by Wehrmacht troops.

The turret armament of this abandoned wz. 34, which in this case was housed in a rectangular mount, has been removed. The large rear door is obvious in this photo. The vehicle had a second door on the left side.

BELGIUM

Although Belgium had developed and built several road tanks during the First World War and employed them between the wars, in part together with French types, no armored cars were on hand when war broke out with Germany.

The *Gendarmerie* still had a number of armored wheeled vehicles of the French Berliet VUDB type, but these were in poor mechanical condition and saw no action.

In 1940 armored prime movers were delivered to the cavalry regiments to serve as tractors for 47-mm anti-tank guns.

They were Ford 91 Y one-ton truck chassis, which were converted to all-wheel drive with the help of Marmon-Herrington components.

The chassis were mated with armored bodies, which were delivered by the Ragheno Firm of Mechelen, at the Ford Works in Antwerp. Production numbers are unknown.

Captured examples of these vehicles, which were actually designed as prime movers, were prized by the Wehrmacht and were employed as armored command vehicles and by reconnaissance units. It is known that they were used by the 14th and 35th Infantry Divisions and by the 8th Panzer Division.

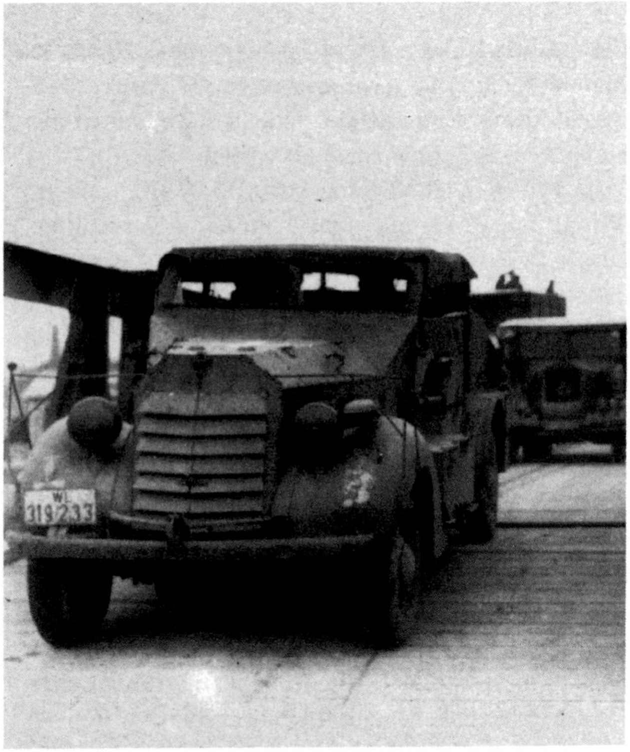
WL
319233

Above: A former Belgian Ford/Marmon-Herrington tractor in service with a unit of the German Air Force (WL-319233). (BA)

Below: This armored tractor was used by the 8th Panzer Division as a command vehicle (WH-754082). In addition to the tactical symbol and German crosses, the vehicle carries the division standard fore and aft. The vehicle was also equipped with radio. (BA)

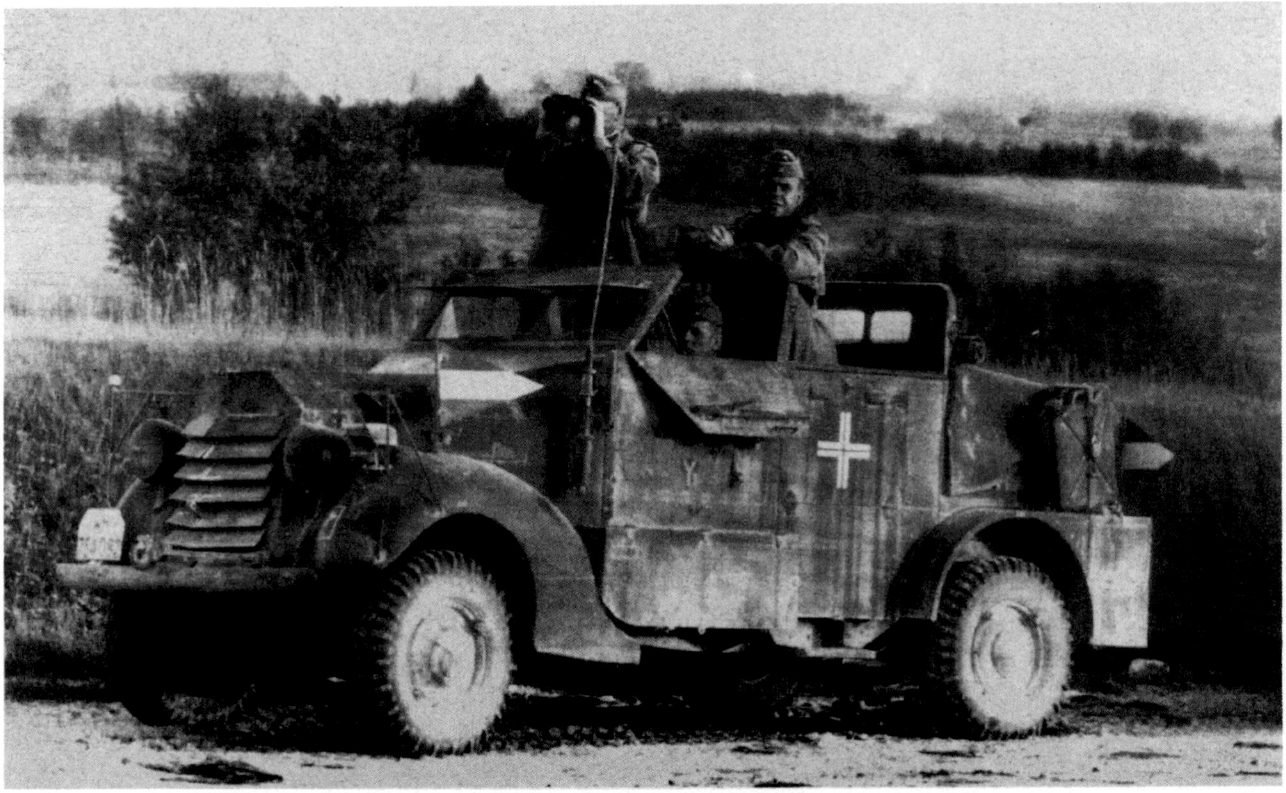

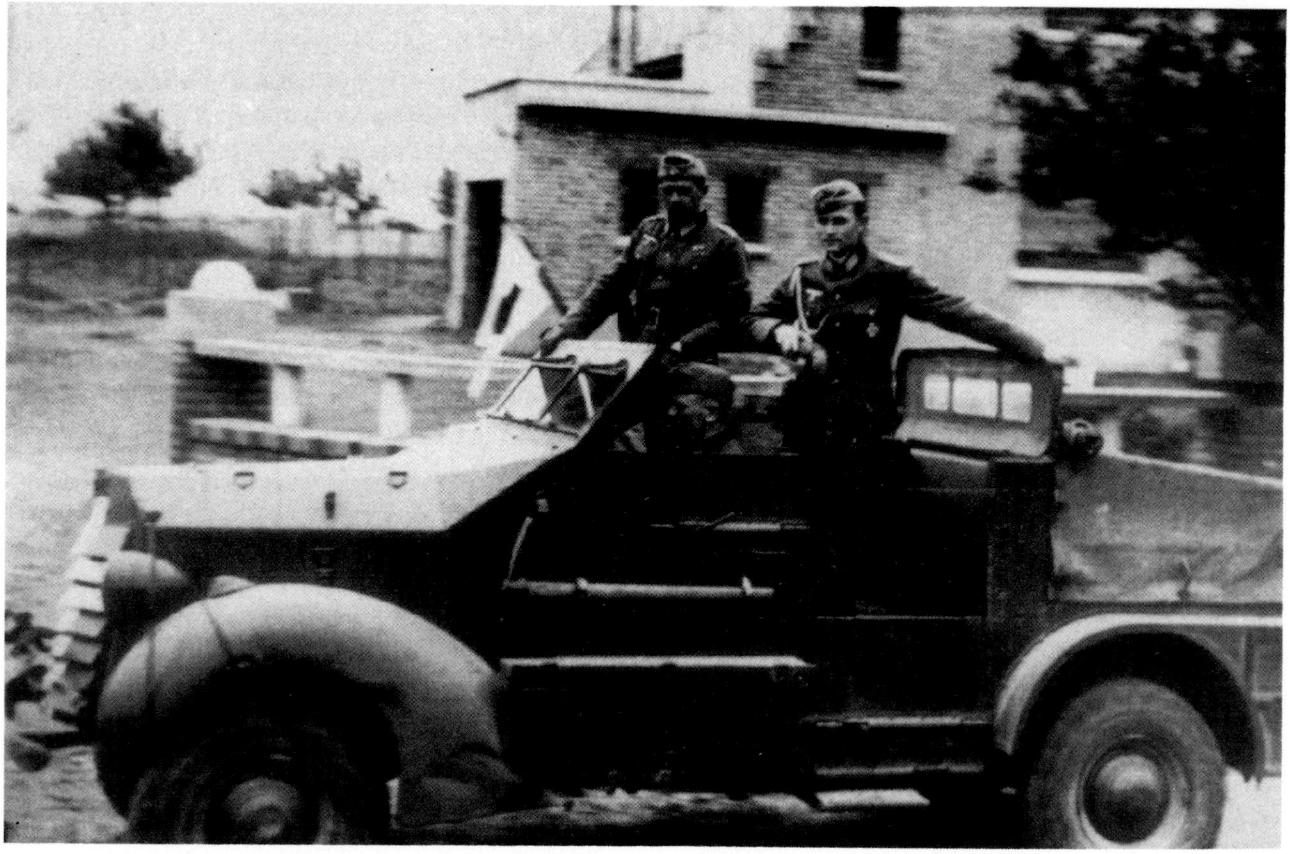

Above: The armored Ford/Marmon-Herrington tractor was also used by I Battalion, 11th Infantry Regiment as a command vehicle. The side armor for the driver and co-driver and the "doors" could be folded down. (RF)

Below: Even an all-wheel drive tractor needed help on roads such as these. This vehicle (WH-446906) of a reconnaissance unit is having problems with the Russian roads. Note the MG 34 mounted in front of the co-driver's seat. (SM)

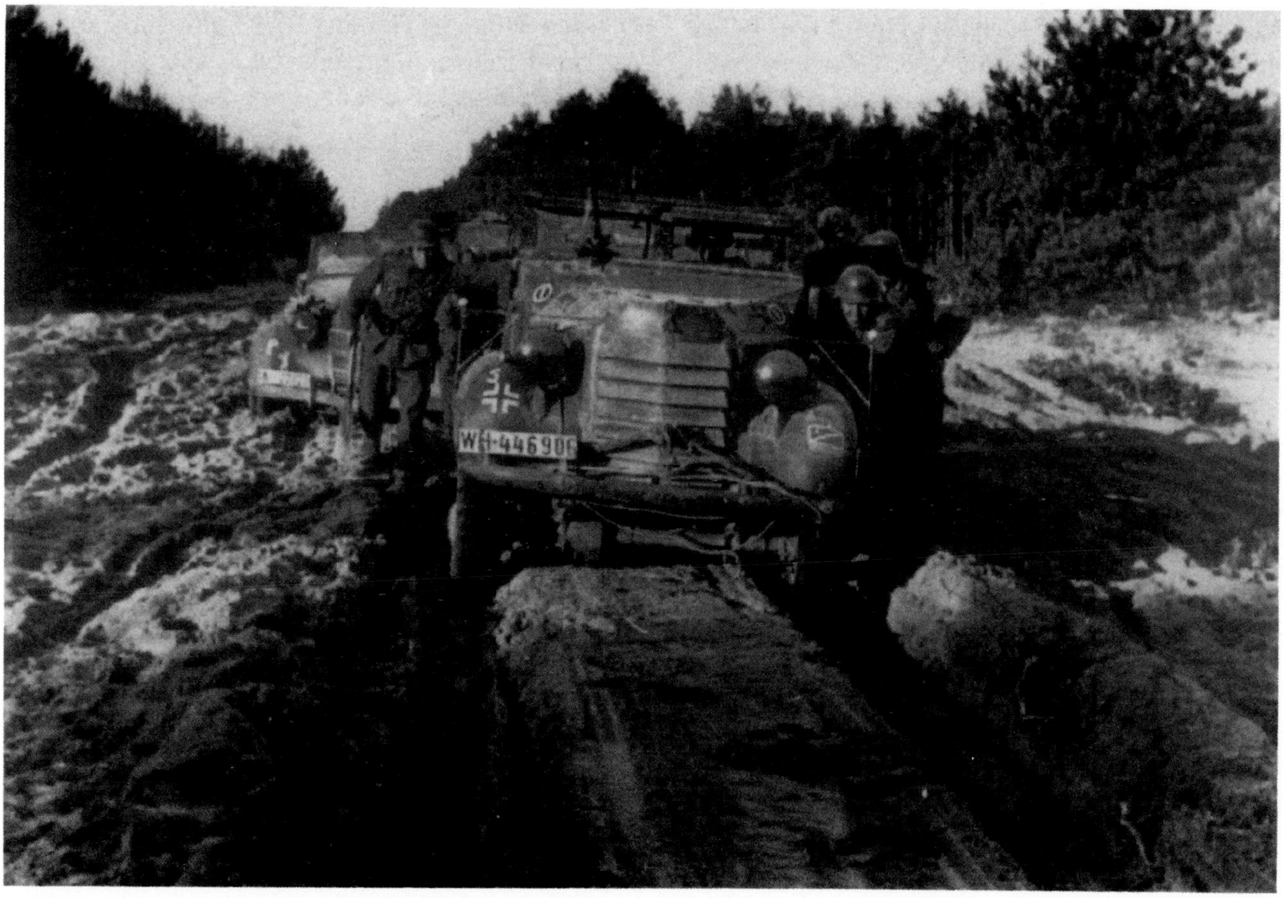

THE NETHERLANDS

The armed forces of the Netherlands had a multitude of different types of armored car and armored wheeled vehicle, however in each case only in small numbers.

After the cease-fire negotiations of the First World War, German troops had handed over an Erhardt truck to the Dutch forces. This vehicle was fitted with armor plate by the firm of Siderius; it remained in service until the middle of the 1930s and in May 1940 was probably still in the depot at Öegstgeest.

In 1929 eight armored cars were built on GMC truck chassis. Five of these vehicles were taken out of service in 1932; the remaining three vehicles were probably still on hand in 1940, but not in action.

In 1932 three armored autos were built on Morris three-axle chassis and each was armed with three or four Lewis machine-guns. These vehicles were also still on hand in 1940.

The above-named armored autos had only limited combat value and those that survived saw no action with German forces. Their fate is unexplained.

The Wilton-Fijenoord (WF) Dock and Shipyard Factory built three wheeled armored vehicles for the Dutch east Indies in 1933. Two vehicles were sold to Brazil, the last vehicle was still in Holland in 1940. The chassis was a Krupp three-axle Type L 2 H 43 chassis with an air-cooled, horizontally-opposed carburetor motor. The 4-5-ton vehicle was armed with three machine-guns and had a crew of three. The armored car was captured in Holland in 1940 and saw its last action in the courtyard of the Reich Chancellery in Berlin, where it was found by Russian troops in 1945.

The Dutch Army formed its first armored car troop in 1936. Twelve Landsverk Type 181 armored cars were procured from Sweden for the troop. Landsverk employed the Daimler-Benz Type G 3 a/P chassis in building these vehicles. The Dutch designated the vehicle, which had a crew of five, "*Pantserwagen M 36*." With a weight of 6.1 tons, the armored car reached a speed of 70 kph. Armor thickness varied between five and nine millimeters. Armament consisted of a Bofors 37-mm gun and a 7.92-mm Lewis machine-gun in the turret, and two more Lewis guns fore and aft in the armored superstructure. The *Pantserwagen M 36* was given the German designation "Panzerspähwagen L 202 (h)." Captured examples are known to have served as armored police cars with the *Ordnungspolizei* (uniformed regular police) and got as far as Russia.

A second Dutch armored car troop was formed in 1938. Once again the contract to supply armored cars was issued to the Swedish firm of Landsverk. This time a Büssing-NAG chassis with a more powerful motor were used. The resulting Landsverk L 180 was designated the "*Pantserwagen M 38*" by the Netherlands. Fourteen M 38 armored cars were delivered, two of them as armored command vehicles with no main armament. The seven-ton vehicle reached a speed of 70 kph. Armor thickness was seven to nine millimeters. Armament was the same as the M 36, the crew consisted of five men. Captured vehicles were used by the German Armed Forces. Two examples were on hand with the Army Ordnance Office's experimental section at Kummersdorf in 1941. Twelve armored cars were ordered from the Van Doorne

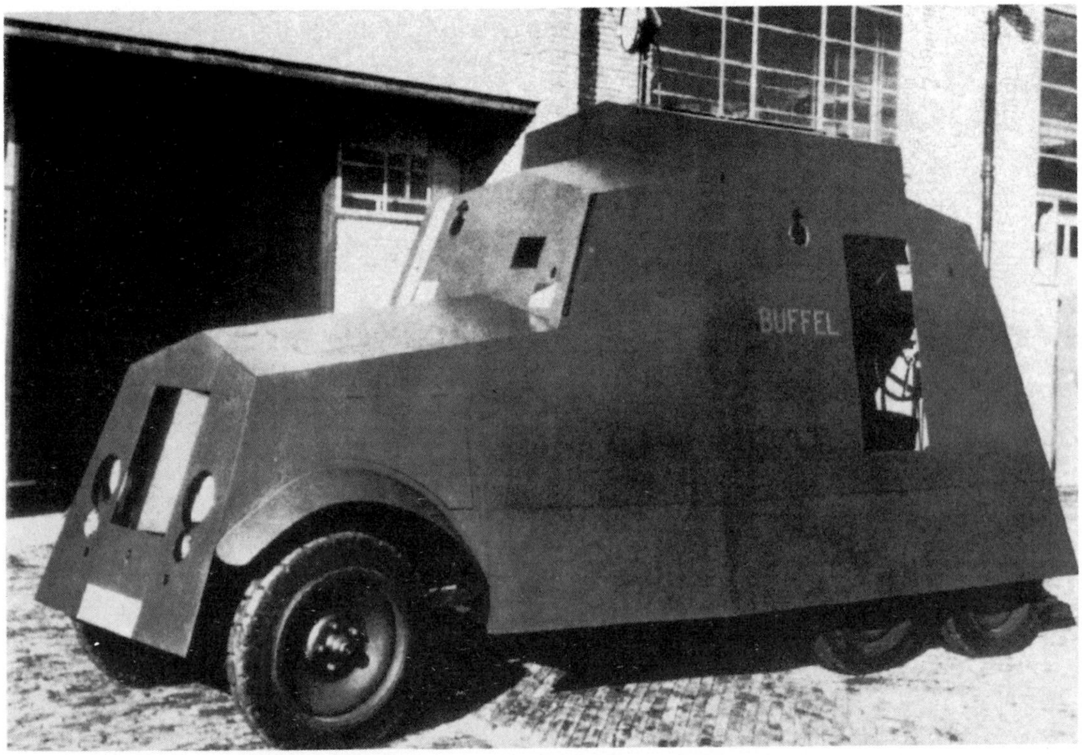
BUFFEL

"BUFFEL" was one of three armored automobiles which were built on Morris three-axle chassis in 1932. This vehicle has mounts for a Lewis machine-gun at the front, rear and sides. The three vehicles, which had front and rear drivers, were probably still in existence in May 1940. Their fate is unknown. (WS)

Aanhangwarenfabriek N.V. (DAF) for the formation of a third armored car troop. The vehicles with the DAF type designation P.T.3 were introduced to service under the name "*Pantserwagen M 39.*" These were very advanced vehicles with integral frames and the DAF-developed Trado drive, which allowed the twin-axled vehicles to become three-axle all-terrain vehicles; thus the type designation P.T. (Pantser with Trado drive). Weighing 4.8 tons, the armored car reached a speed of 75 kph. Maximum armor thickness was 12 millimeters and the crew consisted of five men (including a front and a rear driver). Like the M 36 and M 38, armament consisted of a 37-mm cannon and three machine-guns, even the turret was similar to that of the landsverk types. All twelve vehicles were completed in May 1940, but it is not certain whether all were fully armed. Some of the bulletproof tires were also said to have been lacking. Nevertheless, photos prove that a number of vehicles saw action and were destroyed in combat. Captured M 39 armored cars were assigned the designation "Panzerspähwagen DAF 201 (h)." A captured vehicle was on hand with the Army Ordnance Office's experimental section at Kummersdorf in 1941. From mid-1940 until April 1942 the bicycle battalion of the 227th Infantry Division had a platoon with six Dutch Landsverk armored cars. Since they were tied to the roads, they proved less than successful in the east and were transferred to operational areas where road conditions were more favorable.

Above: In the foreground is a Wilton-Fijenoord wheeled tank on a Krupp chassis, of which three examples were built for the Dutch East Indies in 1933. Behind it is a 1924 Daimler Special Police Car, Type DZVR. The photo was taken in the courtyard of the Reich Chancellery in Berlin in May 1945.

Right:
Two Landsverk Type 181 armored cars during a parade by the Uniformed Regular Police in Holland in February 1941. The Dutch Army introduced twelve of these Pantserwagen M 36 into service in 1936. The armored car was armed with a 37-mm cannon as well as three machine-guns.

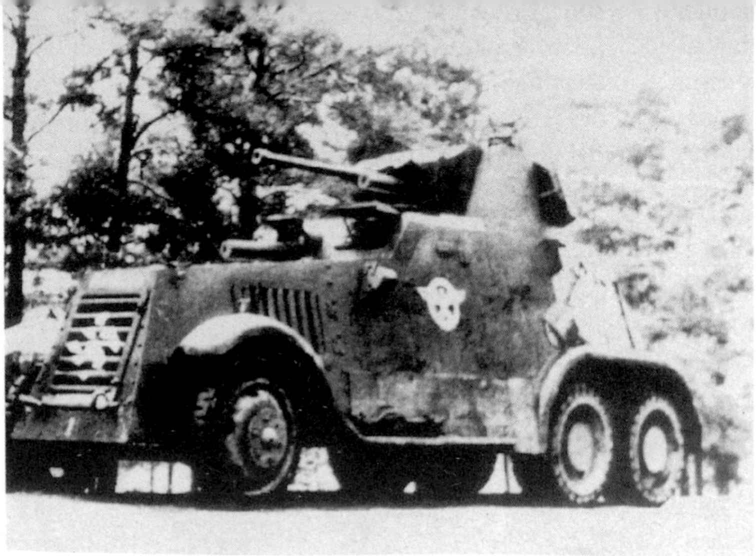

Left:
Several of the Dutch Pantserwagen M 36 served with the Uniformed Regular Police in Russia as the Polizei-Panzerkraftwagen (h). There they were used to patrol occupied territories; this vehicle wears the name "Arnheim."

Right:
This close-up photo shows details of the turret and the driver's visor. Characteristic of these vehicles are the thick jackets of the 7.92-mm Lewis machine-guns. Chains could be mounted on the rear wheels to improve the armored car's cross-country mobility; on this vehicle the chains are stowed on the fenders.

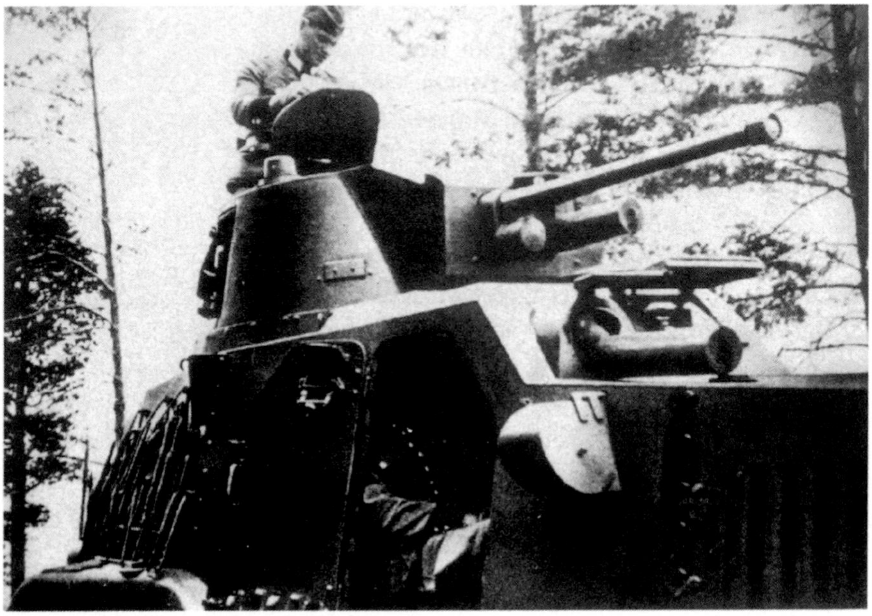

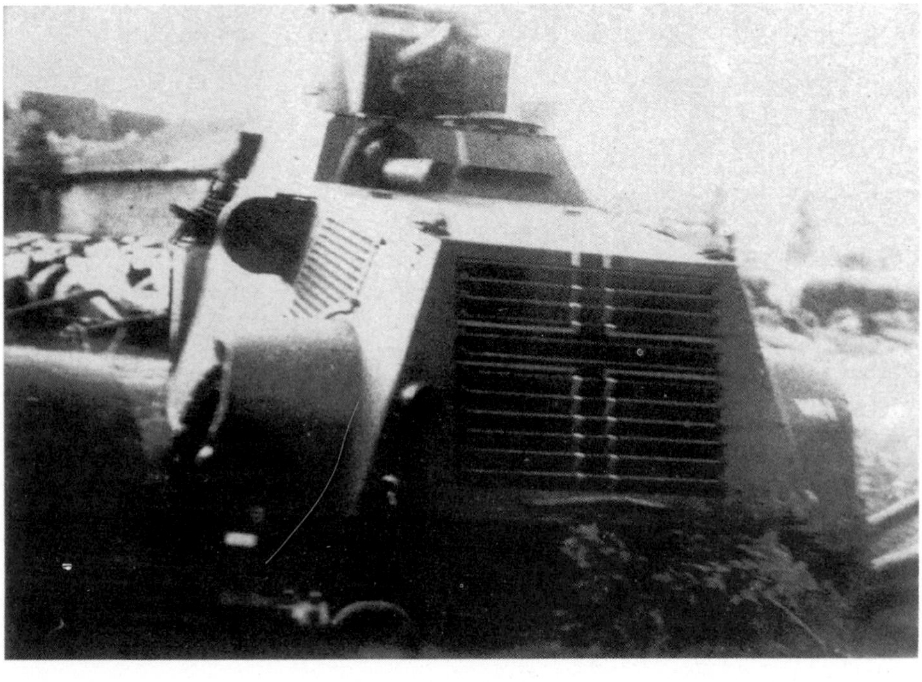

This rather blurry photo shows a Landsverk Type L 180 with German crosses and WH (Wehrmacht-Heer, Armed Forces-Army) number plate. The Pantserwagen M 38 differed from the M 36 in its wider front end with finer radiator ribs and the horizontal cooling vents on the sides of the engine compartment. The M 38 had a more powerful engine than the M 36, armament was the same.

Right:
DAF built twelve examples of the Pantserwagen M 39; however, when war broke out some vehicles had yet to be fitted with armament. The rear of these two vehicles, both of which have had German crosses applied, has no cannon. The vehicles were photographed in the 18th Infantry Division's area in Western Europe in 1940.

Left:
The M 39 had an integral frame and was equipped with the Trado Drive developed by DAF. Auxiliary wheels were mounted on the nose to improve the vehicle's cross-country mobility. Armament remained unchanged from that of the M 36 and M 38 and like the earlier types it had a crew of five.

Right:
The M 39 also saw service in Russia with the German Armed Forces. This Pantserwagen M 39 was lost at the Volkhov in 1941. It is not possible to tell whether the vehicle was knocked out by enemy fire or ra over a mine.

names "Max" and "Moritz" and were used by Rommel and his staff. The third vehicle went to the 21st Panzer Division (General Streich).

Another type of armored car captured by the Germans in North Africa was the Marmon-Herrington, which was manufactured in South Africa by Ford and Dorman Long. In 1938 South Africa built a prototype of a wheeled armored vehicle on a Ford 3-ton truck with rear-wheel drive. A second prototype was fitted with the Marmon-Herrington four-wheel drive and was the subject of extensive testing. 266 Marmon-Herrington armored cars were ordered at the end of 1939; after the blitzkrieg in France the order was increased to 1,000 units, with a planned output of 50 units per week. The vehicle's chassis came from Canada, the Marmon-Herrington four-wheel drive from the USA and its armament from Great Britain. Armor plate was delivered by the South African Iron and Steel Industrial Corp. and final assembly was carried out mainly by Ford and Dorman Long.

The first 113 Mark I Marmon-Herringtons (chassis with rear-wheel drive only) as well as the first Mark IIs (with all-wheel drive) were delivered to South African units and were used mainly against the Italians in East Africa. Like the Mark I, the first Mark II vehicles were equipped with a riveted armored body, however as production of the Mark II went on this was replaced by a welded body.

The first Marmon-Herrington armored cars (all Mark IIs) joined the British forces in North Africa in March 1941. These six-ton vehicles were capable of reaching 80 kph on roads. Armor thickness was 12 millimeters. The armored car's crew consisted of four men. The original armament of the Mark II was one machine-gun in the turret and a second in the body. The British Mark II was fitted with a standard armament of a 14-mm Boys anti-tank rifle and a 7.7-mm Bren machine-gun beside it in the turret. A 7.7-mm Bren gun and a 7.7-mm Vickers gun could also be mounted on the

A recently-captured Marmon-Herrington Mark III armored car. The vehicle's only turret-mounted weapon is a Bren machine-gun; the Boys anti-tank rifle is absent. The Mark III had a more spacious turret than the Mark II.

This Marmon-Herrington Mark II was captured by the same unit; a French 25-mm anti-tank gun has been mounted on the vehicle in place of a turret. It is also armed with a Vickers machine-gun. Various captured weapons, such as the Italian 20-mm and 47-mm guns or the German 37-mm anti-tank gun, were installed in place of the turret in an effort to beef up the armament of the Mark II and Mark III.

turret. The machine-gun on the left side of the body was usually not fitted and the opening was covered with a plate.

A new version, the Mark III, was placed in production in May 1941. This type, too, was based on a Ford chassis, but with a wheelbase that was 45 cm shorter. As well as a number of improvements to the chassis, the Mark III introduced a revised body and a new, larger turret. Armament was similar to that of the Mark II and the body-mounted machine-gun was dropped entirely. 2,630 examples were built by August 1942. The Mark III dispensed with the rear doors of the Mark II; however, a rear door was reintroduced on late-production Mark IIIs as a result of difficulties entering and exiting through the turret and the side doors. In some cases these vehicles were designated Mark IIIA. Other minor modifications were introduced during production of the Mark III.

The subsequent version, the Mark IV, of which 2,116 examples were built, did not take part in the fighting in North Africa. Captured Mark II and Mark III Marmon-Herrington scout cars were used by the German Armed Forces in North Africa as armored cars or as armored observation vehicles by the artillery.

Below: Captured Daimler Dingo Mark II armored car, photographed on 13 April 1942 near Chechiban, Libya. The car, which bears the name "Purzel", has the number code WH-733549, which suggests that it has been in Wehrmacht service for some time. Note the revised radiator grille compared to that of the Mark I. (BA)

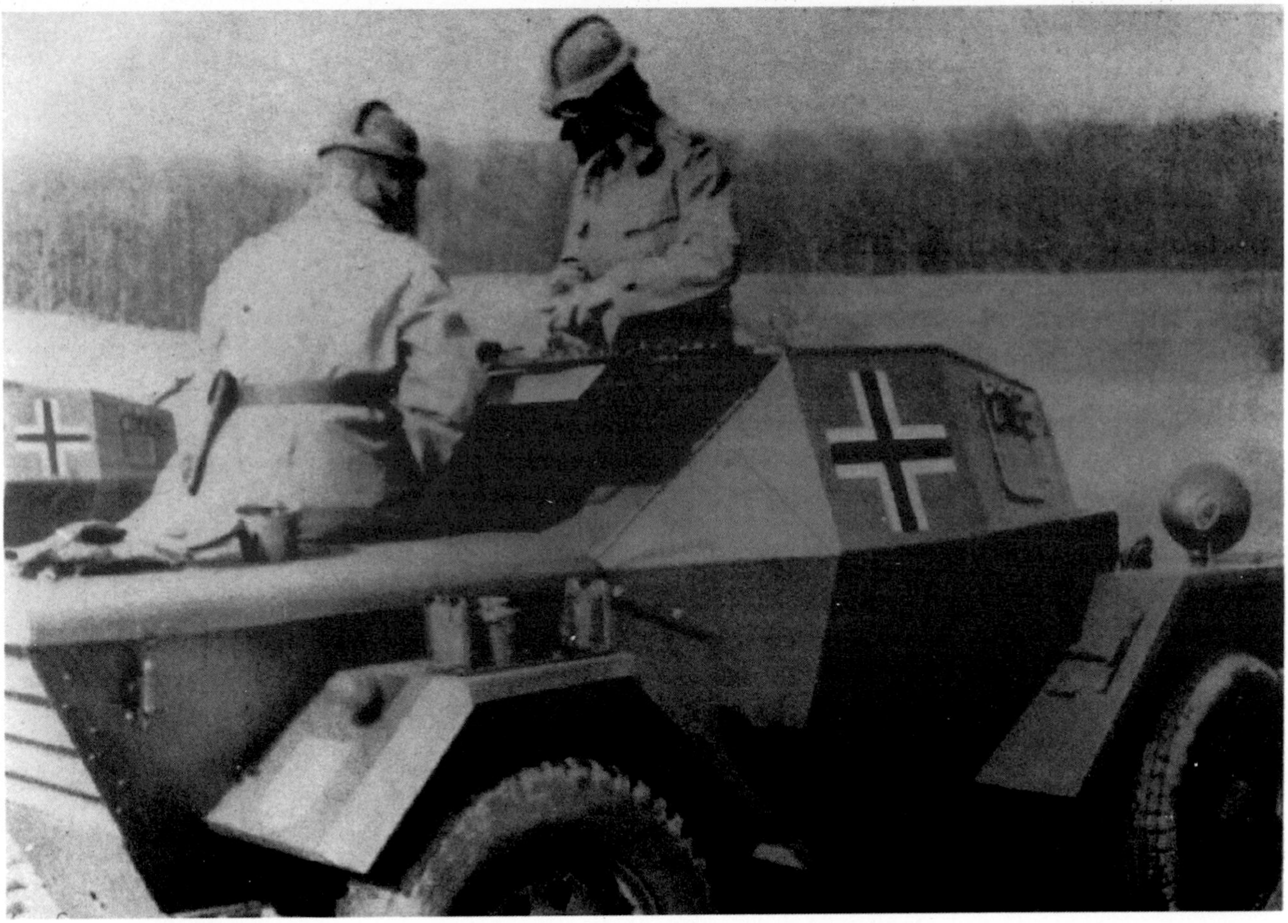

Above and below: A Daimler Dingo Mark I of the 1st Navy Combat Vehicle Battalion in the Black Sea region. The armored cars were armed with a MG 34 machine-gun and were used to escort transport columns. The crews are equipped with protective helmets of the type formerly used by the Reichswehr.

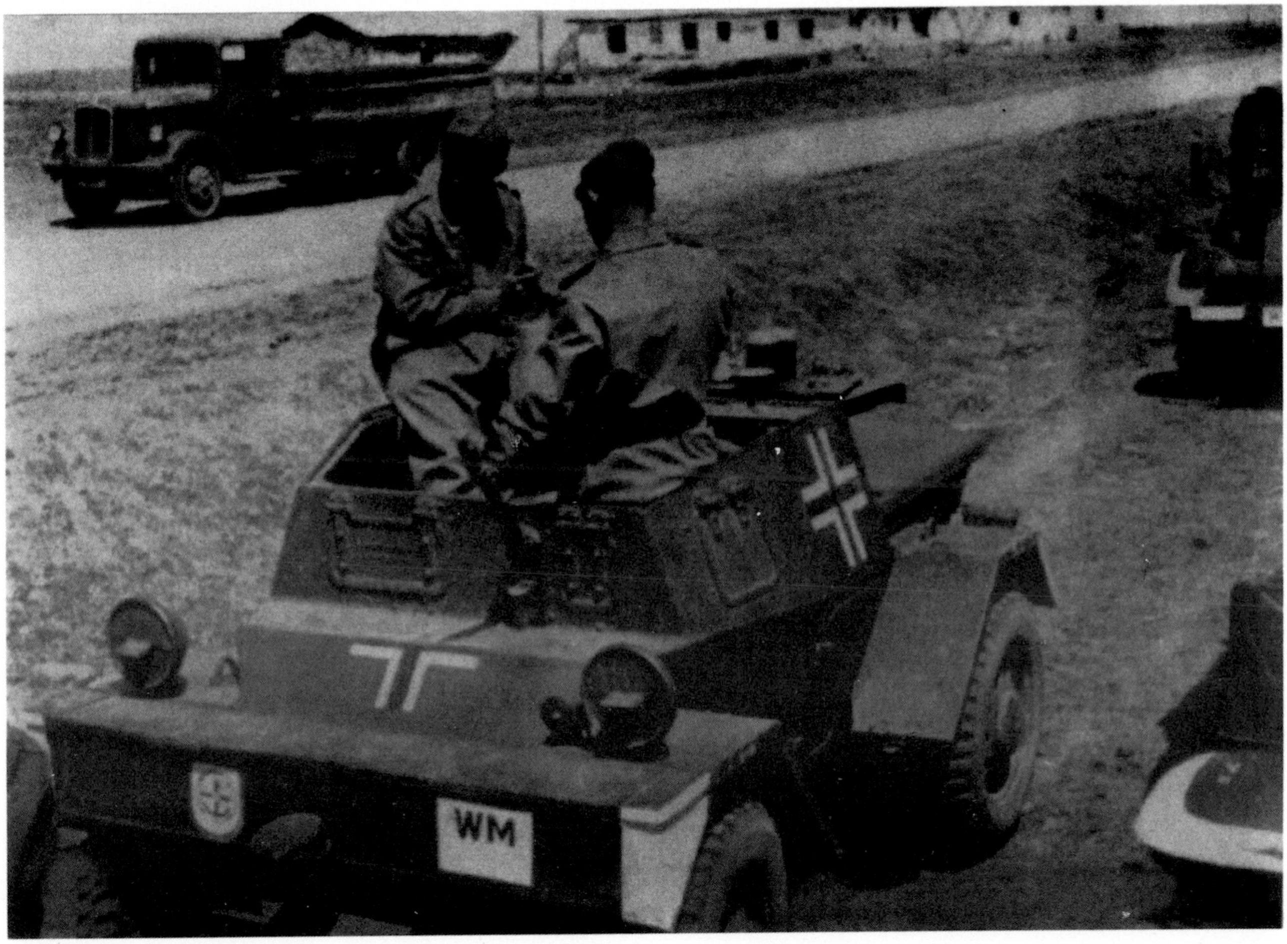
WM

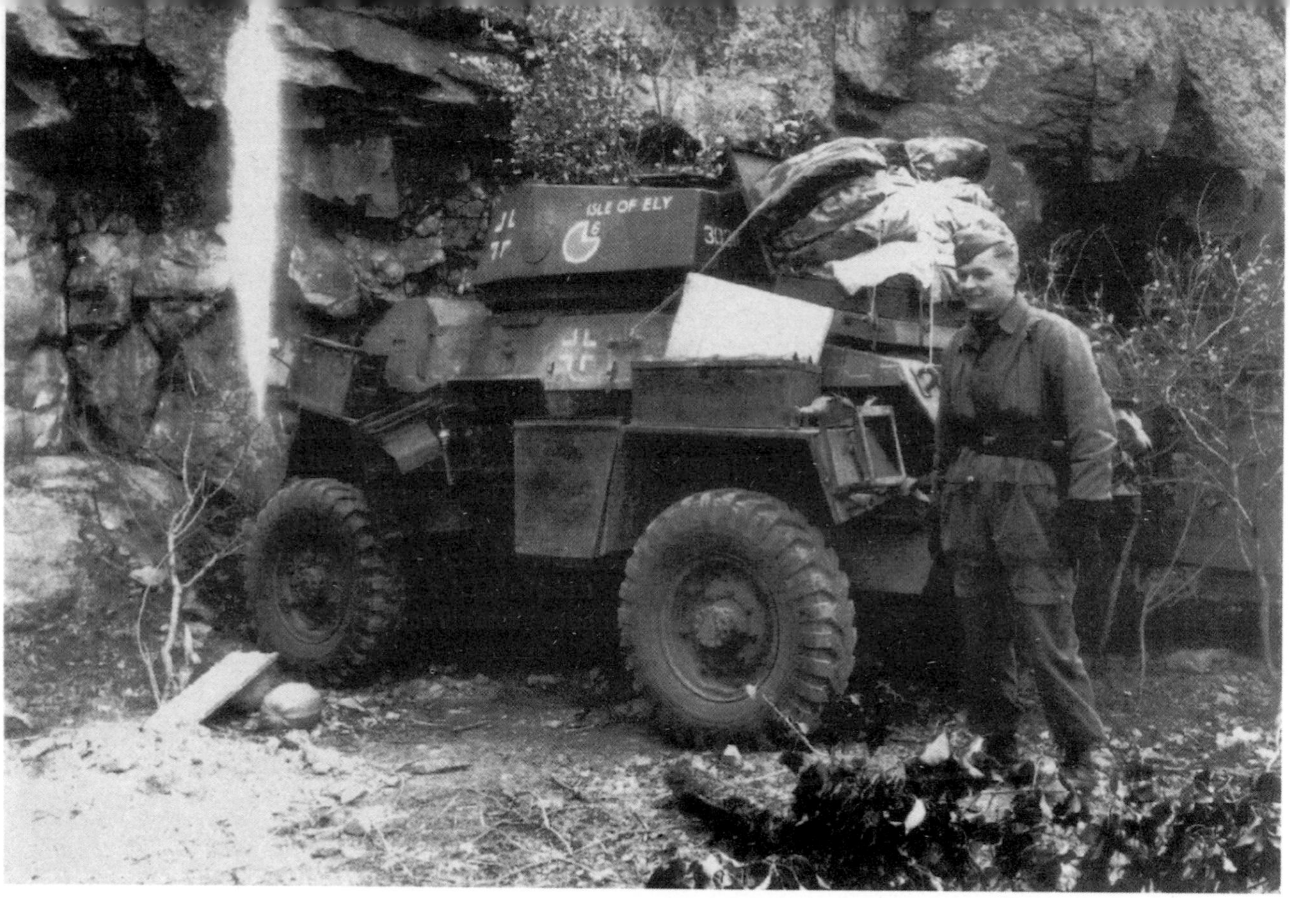
ISLE OF ELY

Above: Humber Mark III "Isle of Ely" in use by the 4th Parachute Division in Italy. The unarmed armored car is marked with German crosses and packed with items of equipment. (BA)

Below: Humber Mark I with the name "Hans" on the left side of the turret in Africa in 1942. The driver's compartment of the Humber Mark I was not an integral part of the superstructure. The vehicle formerly belonged to the armored reconnaissance regiment (12th Lancers) of the British 1st Armoured Division. (BA)

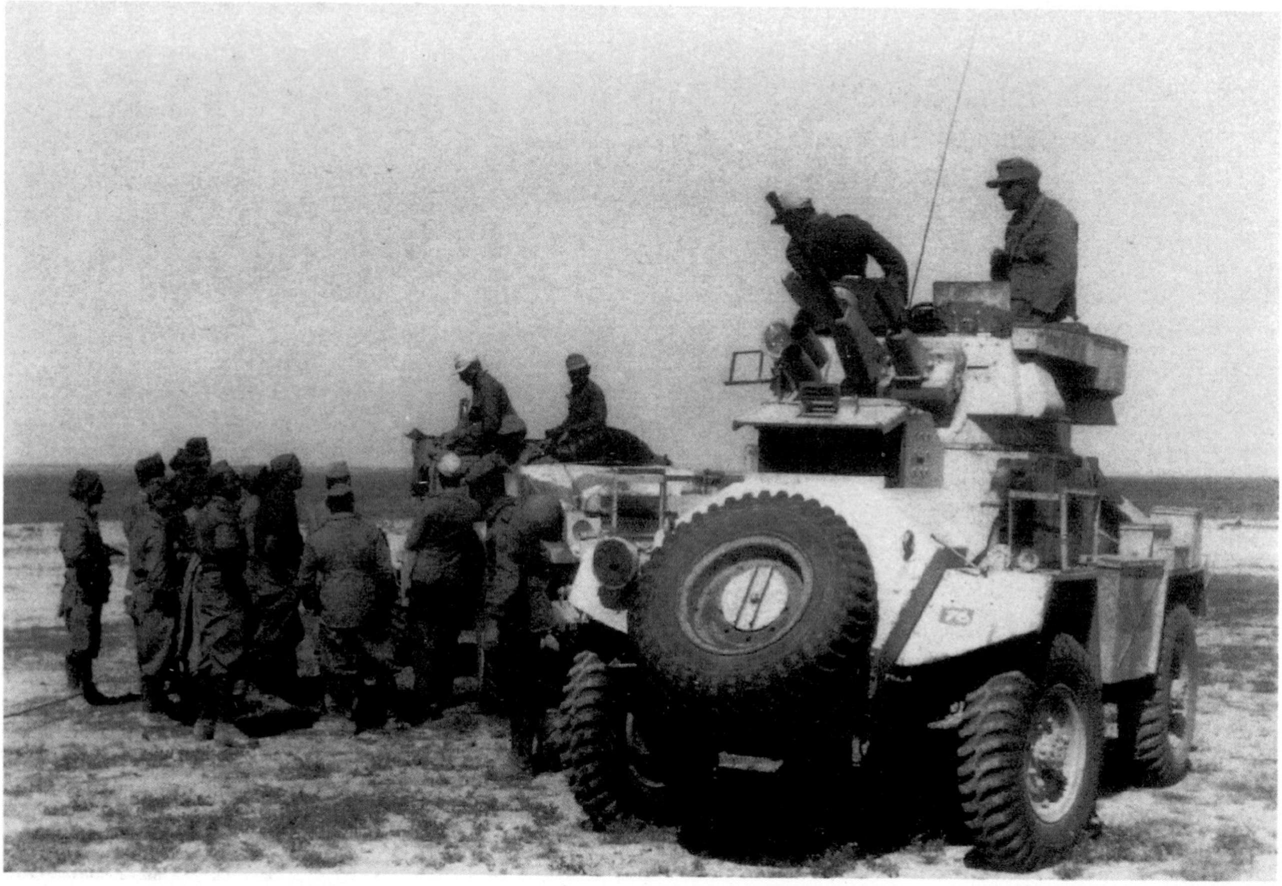

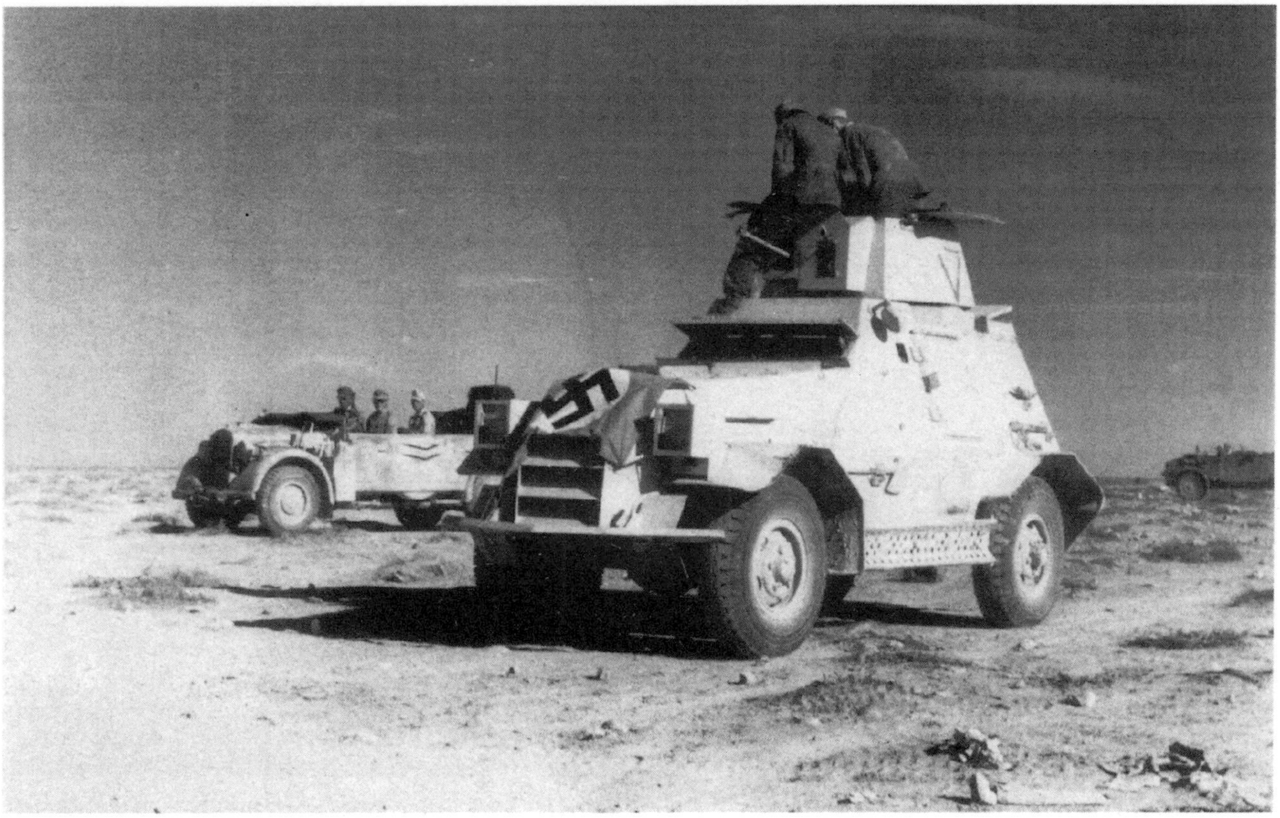

Above: Marmon-Herrington Mark III with white crosses and a swastika flag on the hood as an air identification panel. The car's armored radiator grille identifies it as an early Mark III. The armored rectangular headlights were typical features of the Mark III and were deleted from the Mark IIIA. (BA)

Below: Marmon-Herrington Mark III, late version, with fully-armored radiator in Tunisia in 1943. This vehicle, which served as a model for the title illustration, was used as an observation vehicle by the 4th Battery, 155th Armored Artillery Regiment, a unit of the 21st Panzer Division. (BA)

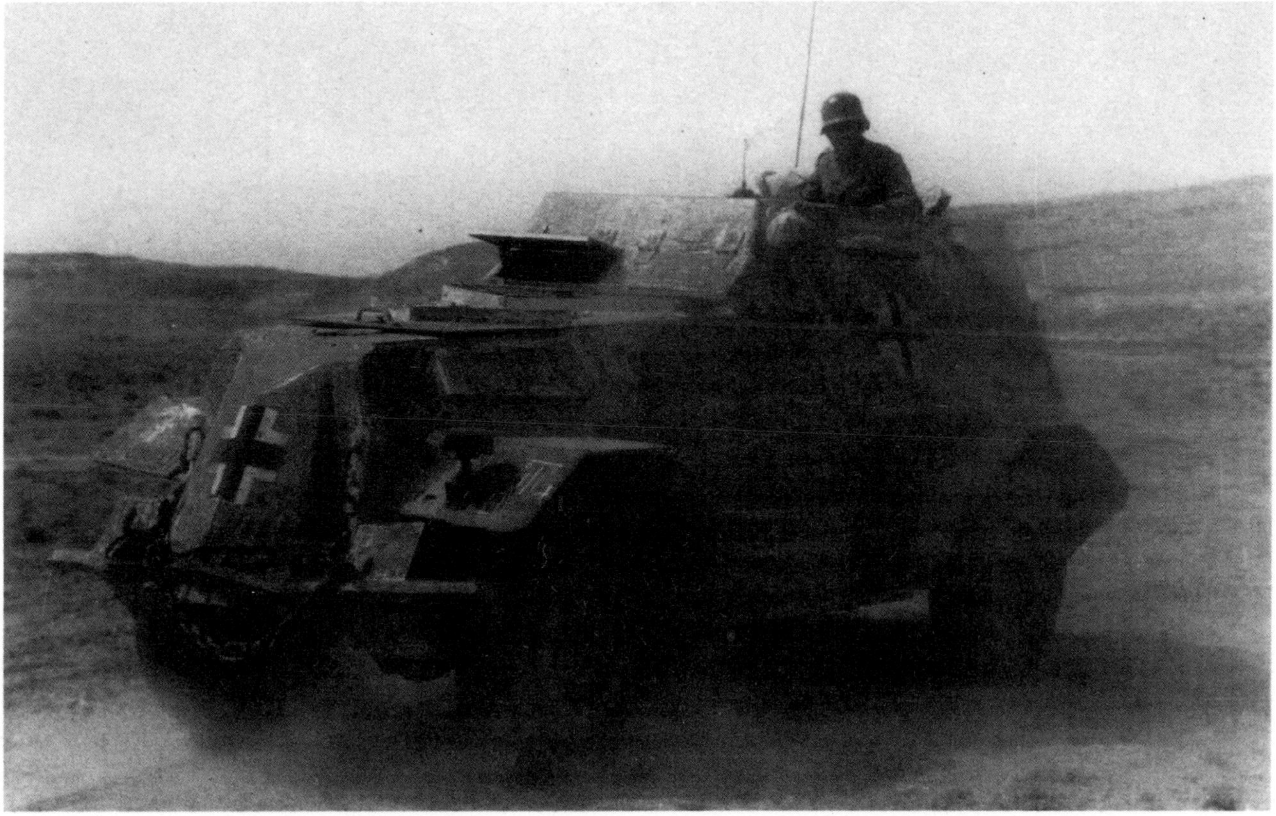

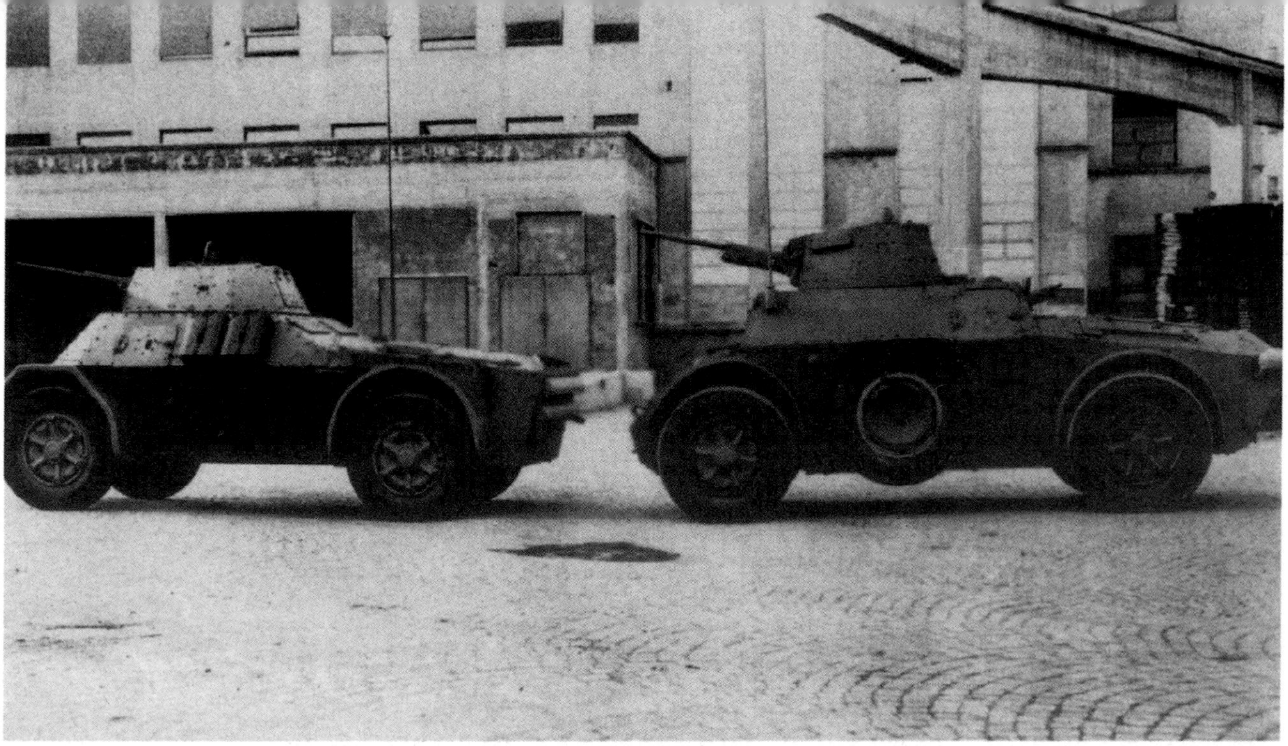

Above and below: The AB 41 armored car (right) was supposed to have been replaced by the AB 43 (left) in 1943. The prototype of the new model had a revised body in which the auxiliary wheels were deleted and the turret moved forward. The rear machine-gun was deleted and the radiator grille was altered. The vehicle received a new turret which was shallower but roomier and which could accommodate either a 20-mm or 47-mm cannon. (BA)

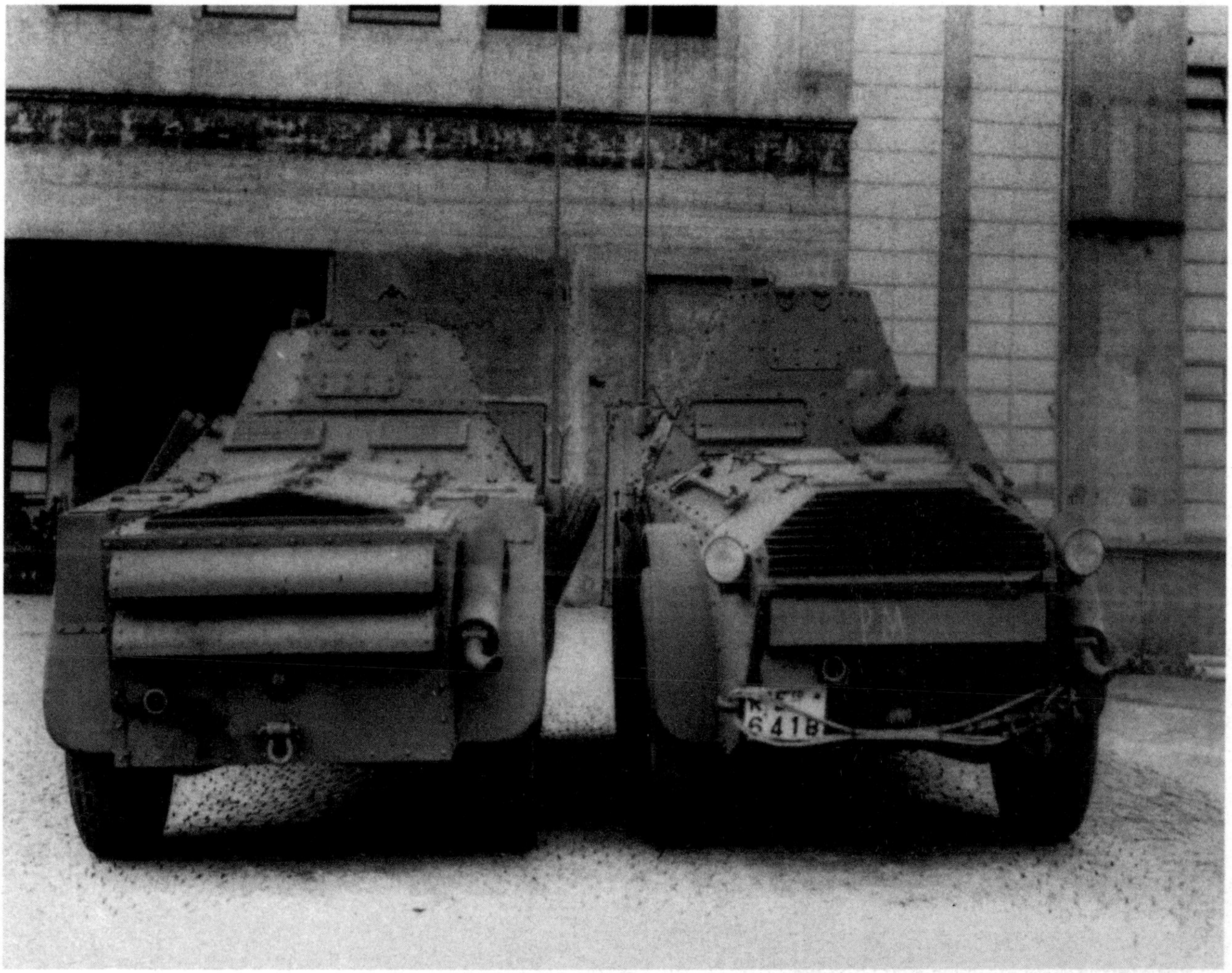

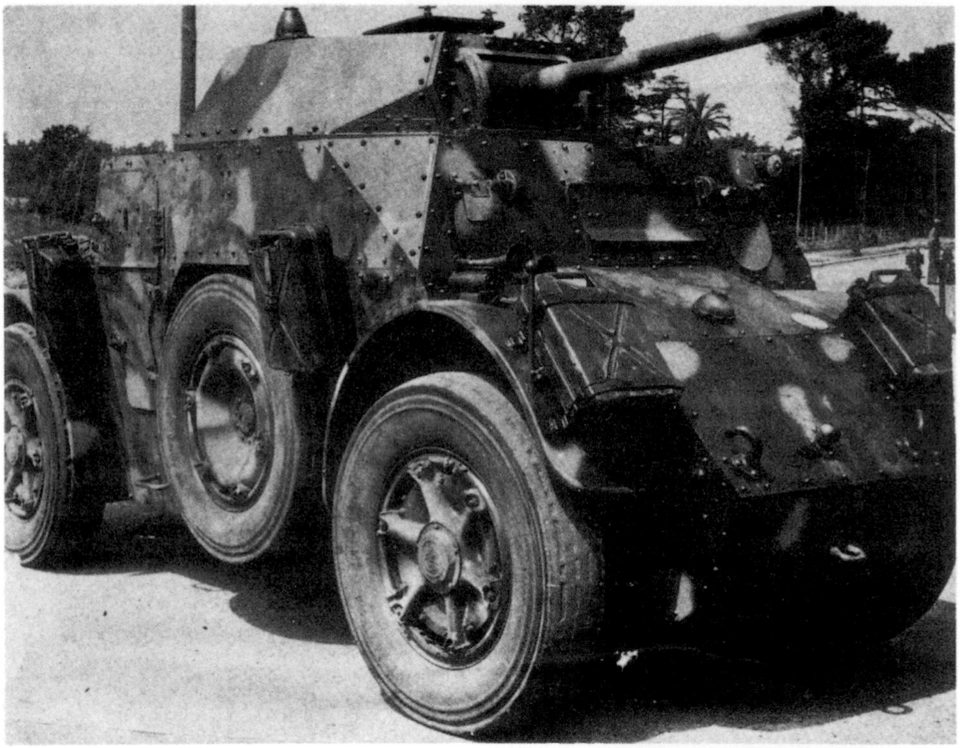

Left:
A prototype of the AB 43 equipped with a 47-mm L/32 tank cannon existed in September 1943. The vehicle was in fact an AB 41 with a widened body and the new turret. (WS)

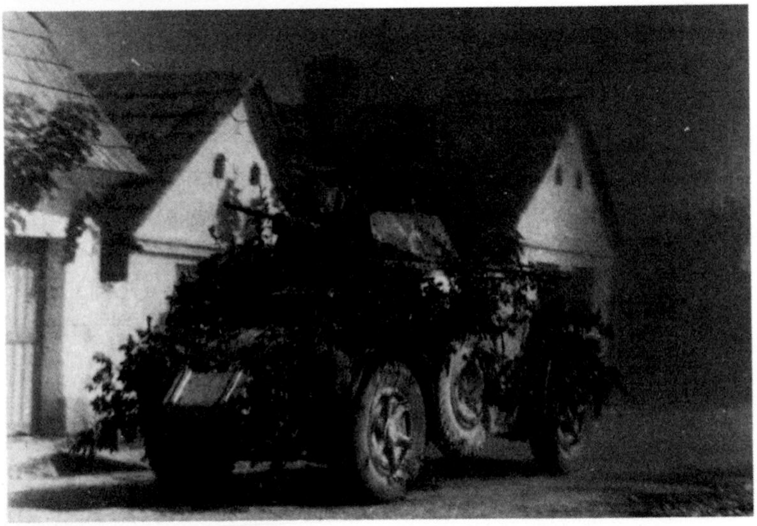

Right:
The vehicles ultimately built under the designation AB 43, however, were unmodified AB 41s which received only the new turret. These armored cars were often designated AB 41/43. This well-camouflaged vehicle was photographed in action in Hungary in October 1944. (BA)

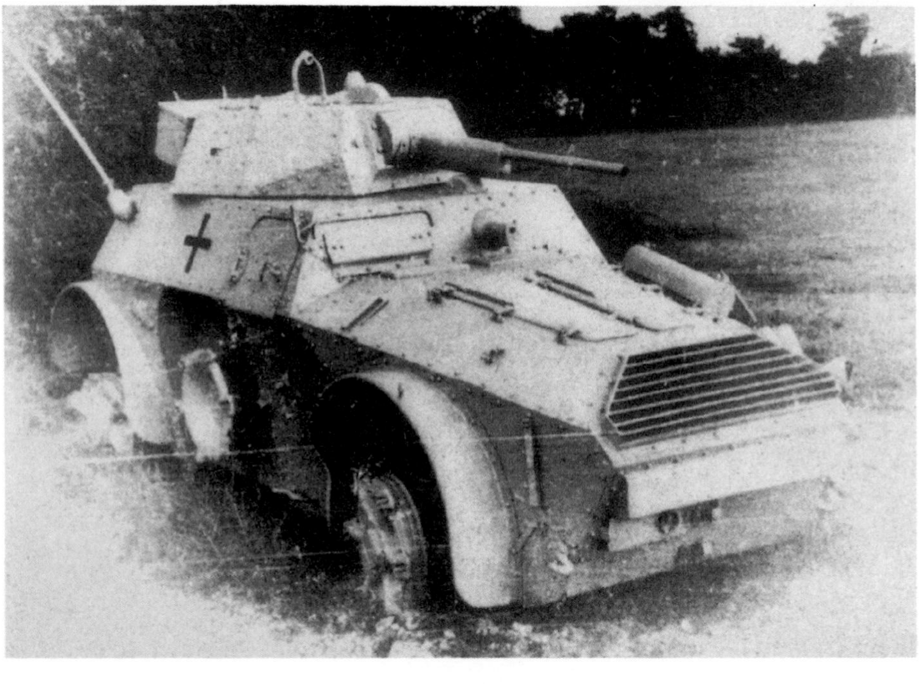

Left:
This AB 41 was also equipped with the new turret and the 20-mm cannon. The vehicle has lost all its tires and the rear-mounted machine-gun has been removed. The storage bin on the rear of the turret is interesting. (HH)

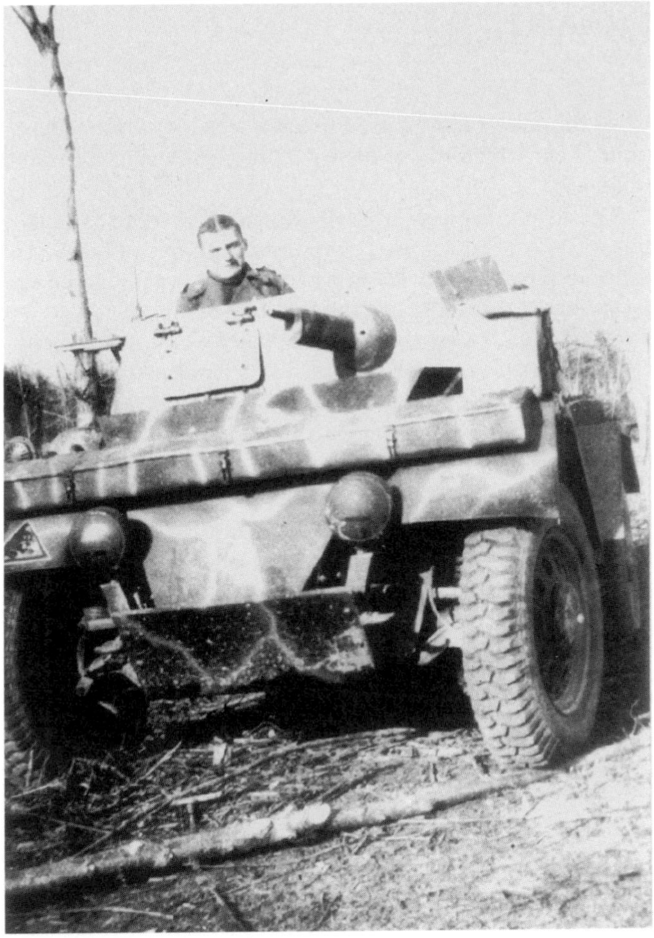

The Italian Army decided in 1943 to build a copy of the English Daimler Dingo for use as a light armored car. The vehicle was armed with a single machine-gun. (JW)

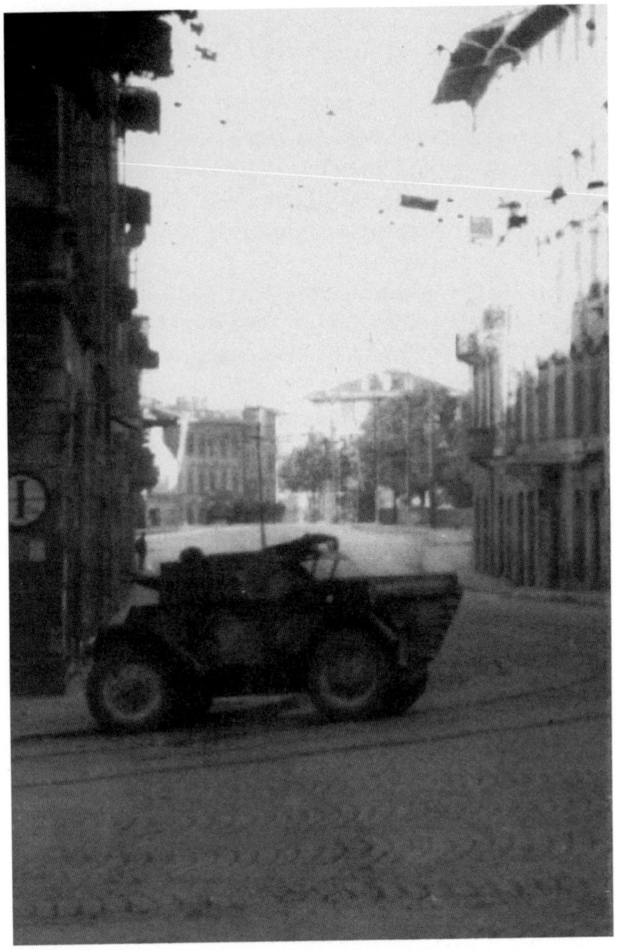

Built by lancia, the armored car was called the "Lince" and at first glance the only difference from the Daimler Dingo was the machine-gun. Here a vehicle of a parachute unit during street fighting in an Italian city. (BA)

This rather blurry photo shows the same vehicle as above left in action with I Battalion, 4th Panzer Regiment. The Wehrmacht issued a contract for the construction of 300 of these vehicles in 1943; more than 100 were delivered by the end of the war. The "Lince" was used as a scout car by reconnaissance units and as a liaison vehicle by unit headquarters. (JW)

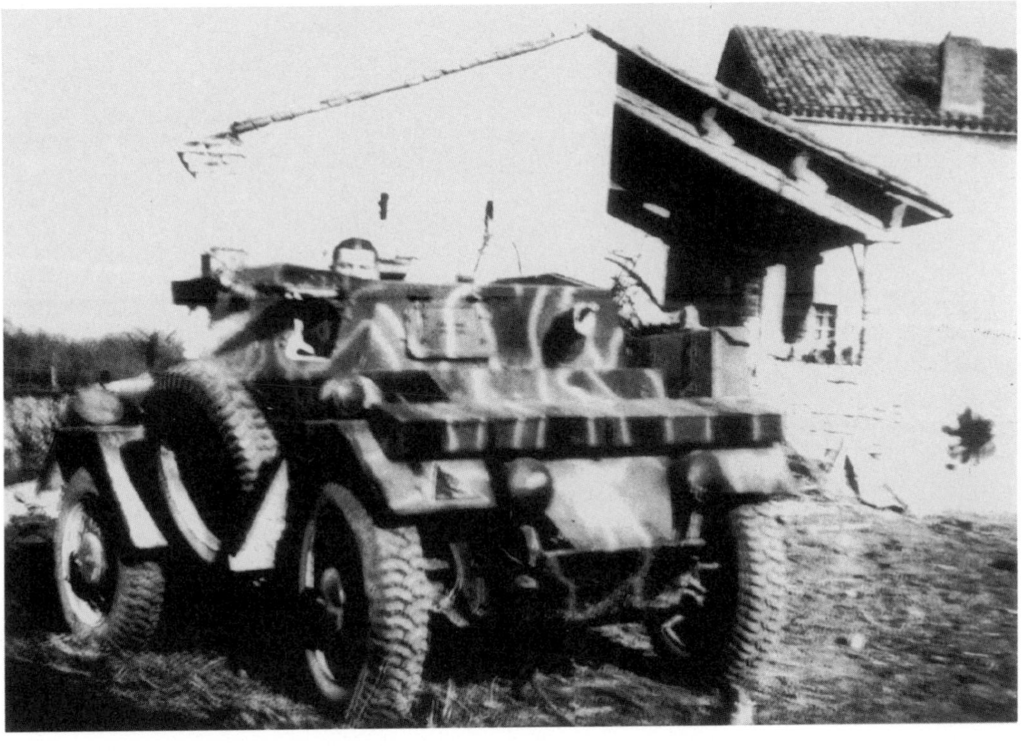

UNITED STATES OF AMERICA

Armored motor vehicles were built and tested in the United States during the First World War and in the period between the wars. Great attention was paid to the manufacture of standardized army vehicles in order to simplify deliveries of spare parts.

In 1938 the White M3 Scout car was accepted as a scout car; it was followed in 1939 by the ultimate version, the improved M3A1. It was a twin-axle, open-topped vehicle with a forward body typical of some half-tracked vehicles.

The 5.6-ton vehicle was capable of a maximum speed of 88 kph.

The M3A1 could accept a crew of eight and its main armament was usually a 12.7-mm machine-gun.

21,000 of these vehicles were built and examples were delivered to Great Britain and Russia under the Lend-Lease Agreement.

The German designation was "Panzerspähwagen M3A1 210 (a)." There is no evidence that captured vehicles of this type were used by the German side.

At the end of 1941 a specification was issued for a light armored car which was to have a 37-mm cannon as its main armament.

These vehicles were actually intended to serve as tank-destroyers. A Ford Motor Company design for a three-axle vehicle designated T22 E2 was accepted; it was introduced into service in June 1942 as the M8 light armored car.

Weighing 7.7 tons, the vehicle reached a maximum speed of 88 kph and was armored from 3 to 19 millimeters.

The M3A1 had a crew of four. A 37-mm M6 cannon and a co-axial 7.62-mm machine-gun were installed in an open-topped turret.

In some cases a 12.7-mm anti-aircraft machine-gun was also mounted on the turret. By the end of 1944 Ford built 8,523 examples of the M8 armored car, which was also called the "Greyhound." The vehicle was used by the Wehrmacht under the designation "Panzerspähwagen M8 (a)." The 11th Armored Reconnaissance Battalion, for example, included two M8s in its vehicle complement on 11 February 1945.

"Sweet Sue", an M8 armored car of the American 3rd Army, was captured in Northern France in autumn 1944. The car lacks fenders, which was typical of these vehicles. A single-axle trailer is still coupled to the armored car. The M8 "Greyhound" was a very modern vehicle, which served with the police and federal border guards after the war. (BA)

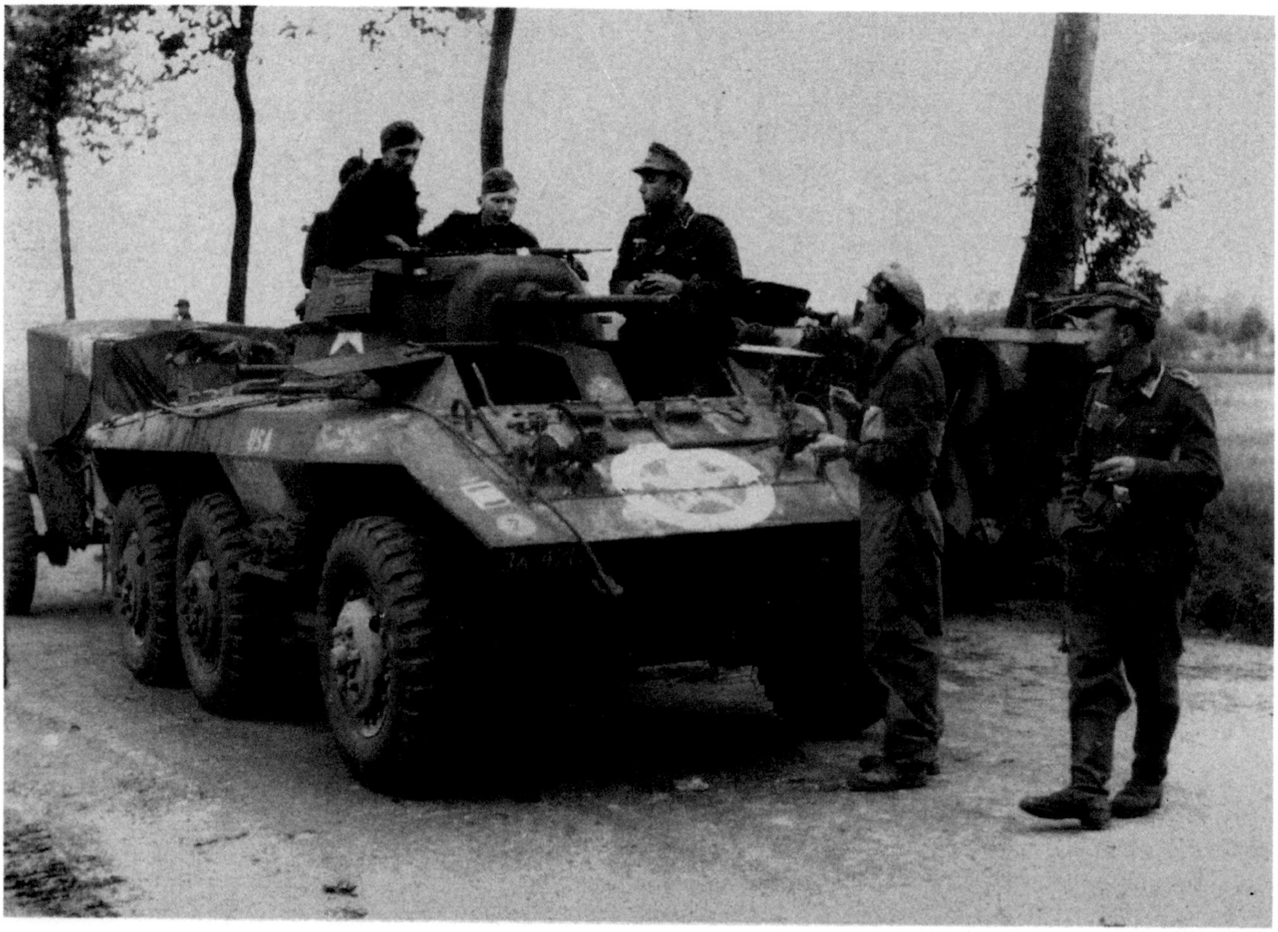
USA

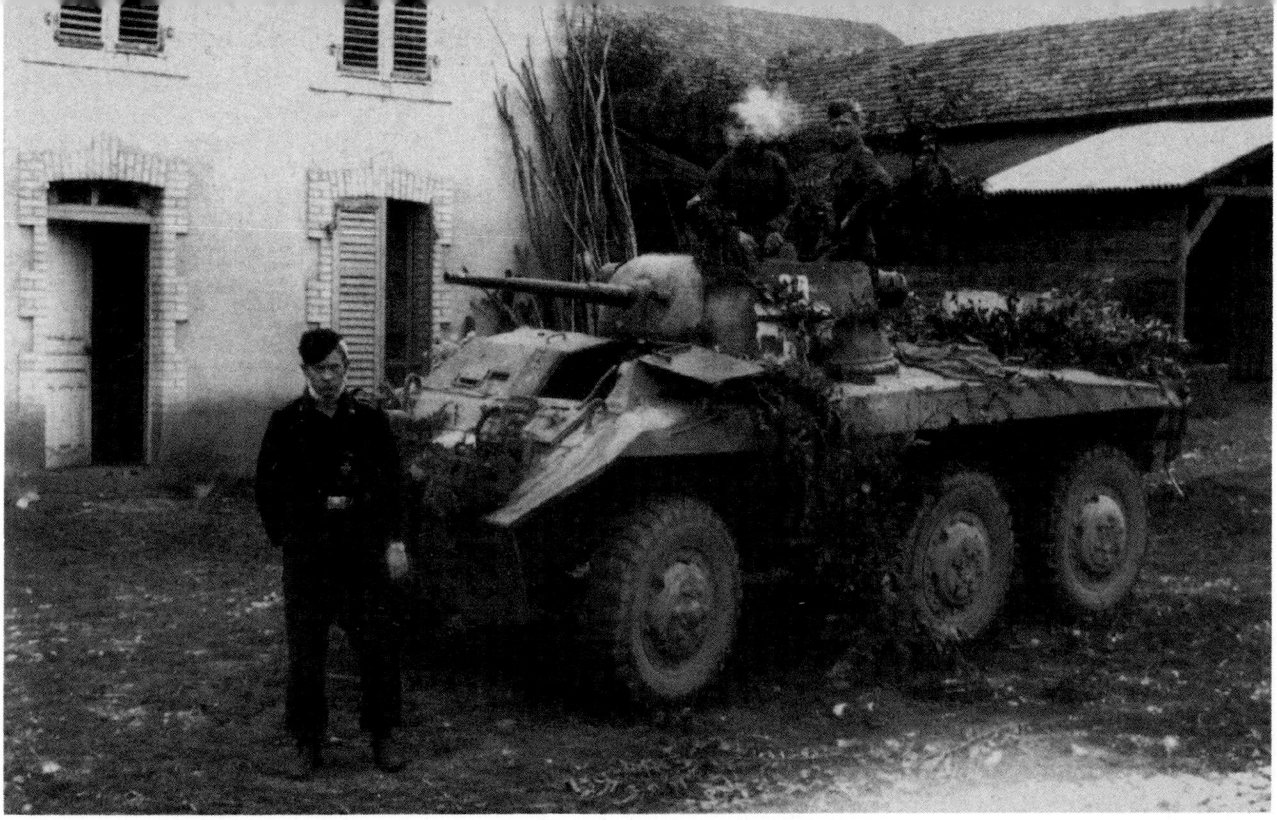

"Sweet Sue" with its new owners. Almost all American markings have been obliterated. The Wehrmacht crosses have been very hastily applied, causing the paint to run. (BA)

Below: A second M8 from the same unit. This vehicle still has its fenders and an anti-aircraft machine-gun is mounted on the turret. The crew's things were carried on the vehicle in duffle bags and boxes. (BA)

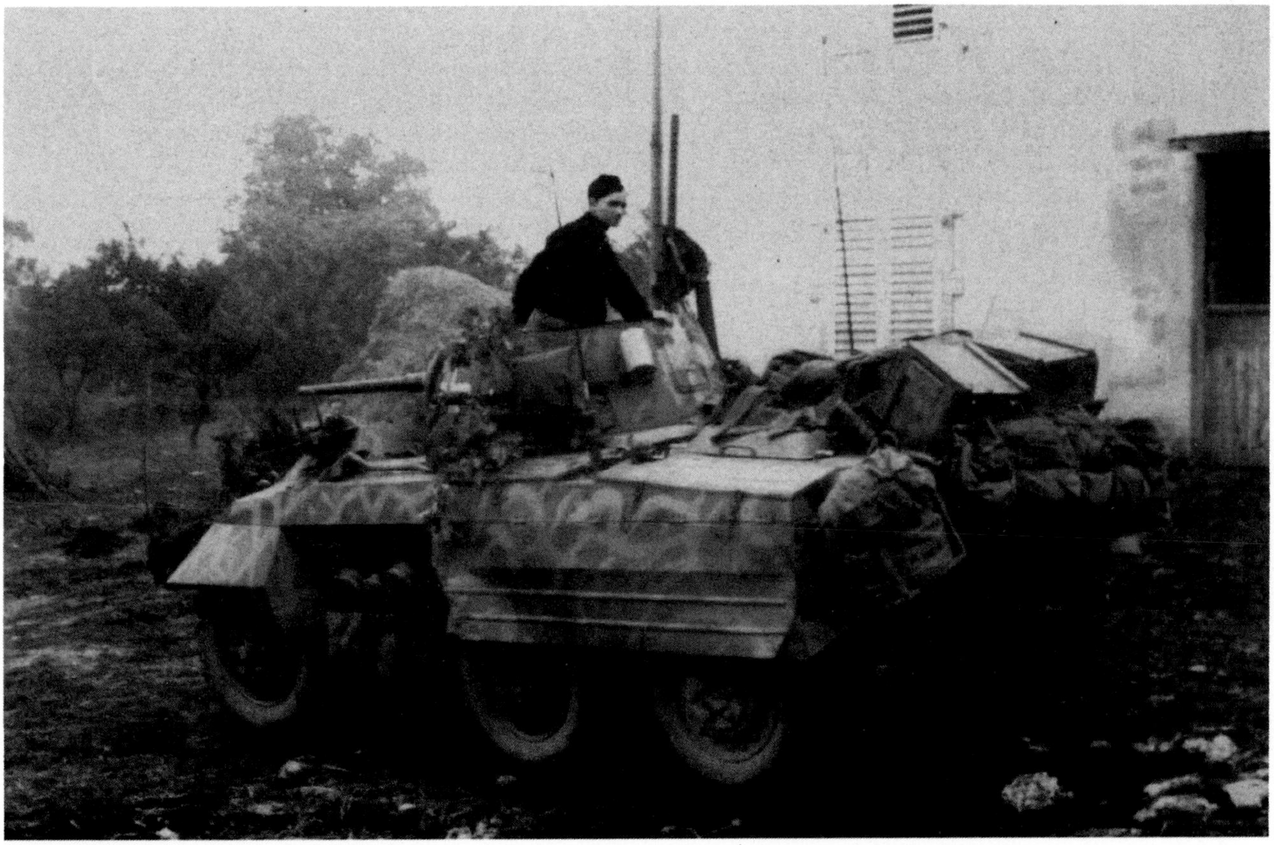

Spanky used the money that his daddy gave him and he bought himself a casino.
Everybody knew that, at casinos, the House ALWAYS wins.
But, after a few years, the casino closed
and Spanky went bankrupt. In fact, he went bankrupt 5 TIMES!

Spanky was bad at business.

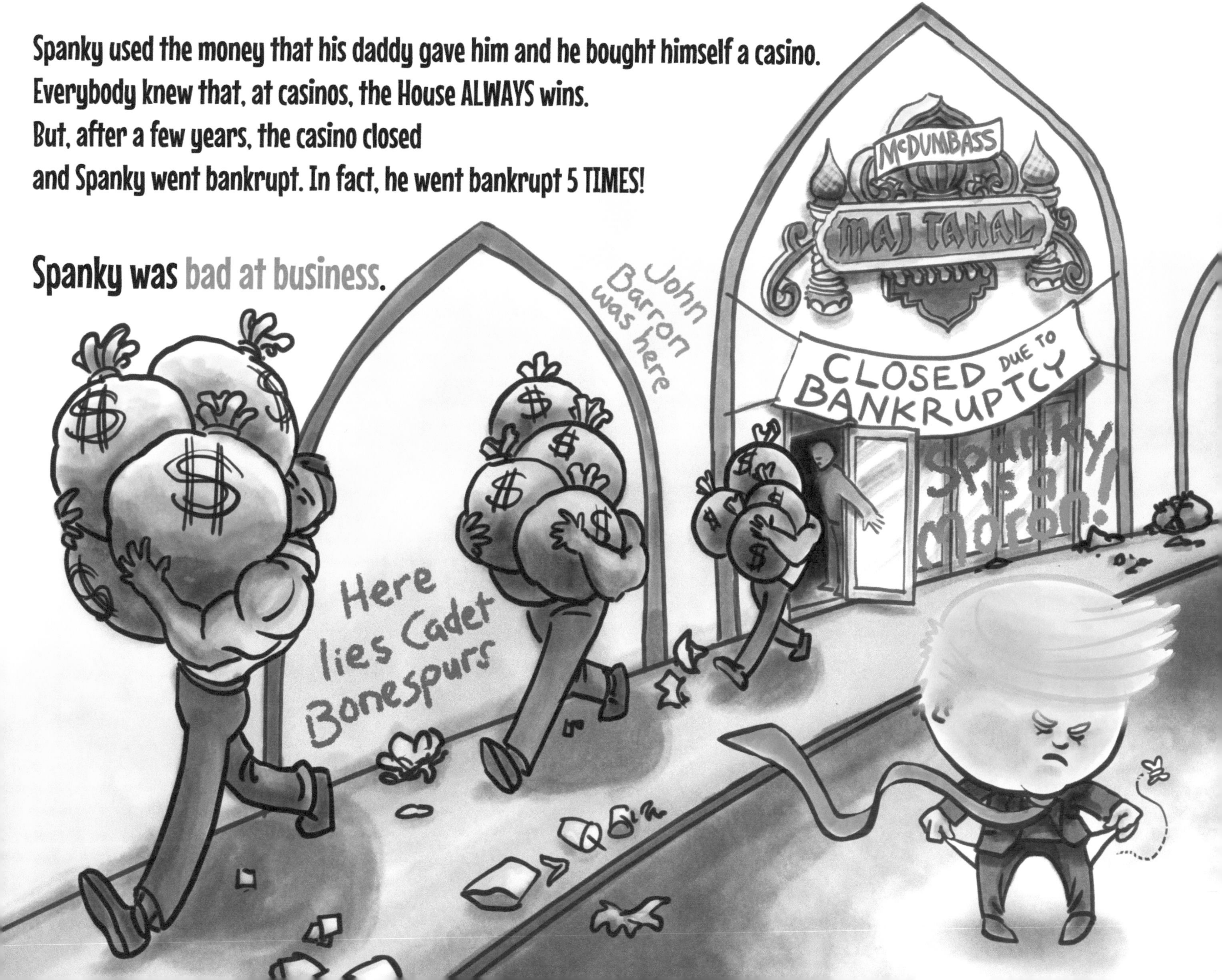
McDUMBASS
MAJ TAHAL
CLOSED DUE TO BANKRUPTCY
Spanky is a moron!
John Barron was here
Here lies Cadet Bonespurs

Spanky met a woman, Eyebonka, and married her. They had three kids: Spanky Jr, Blanky, and his daughter Stanky.

Spanky was bad at marriage.

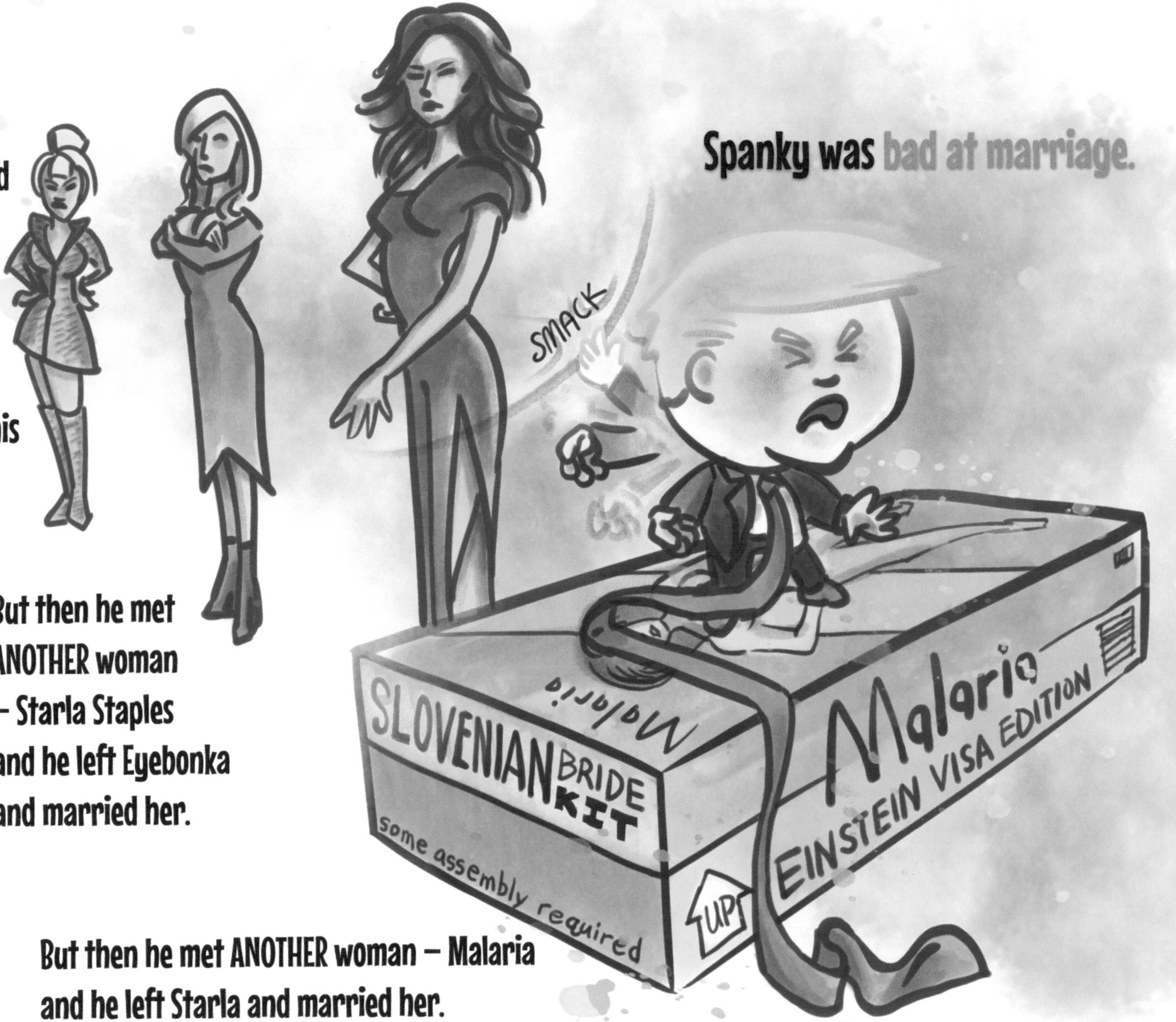
SMACK
SLOVENIAN BRIDE KIT
some assembly required
UP
Malaria
Malaria
EINSTEIN VISA EDITION

But then he met ANOTHER woman – Starla Staples and he left Eyebonka and married her.

But then he met ANOTHER woman – Malaria and he left Starla and married her.

Despite having many, many wives
Spanky had a strange fixation on his own daughter, Stanky.
He often talked about her as a sex object.
He often took weird,
inappropriate pictures with her.
He even told people that
if she weren't his daughter
He'd be dating her.
Spanky was a bad, creepy dad.

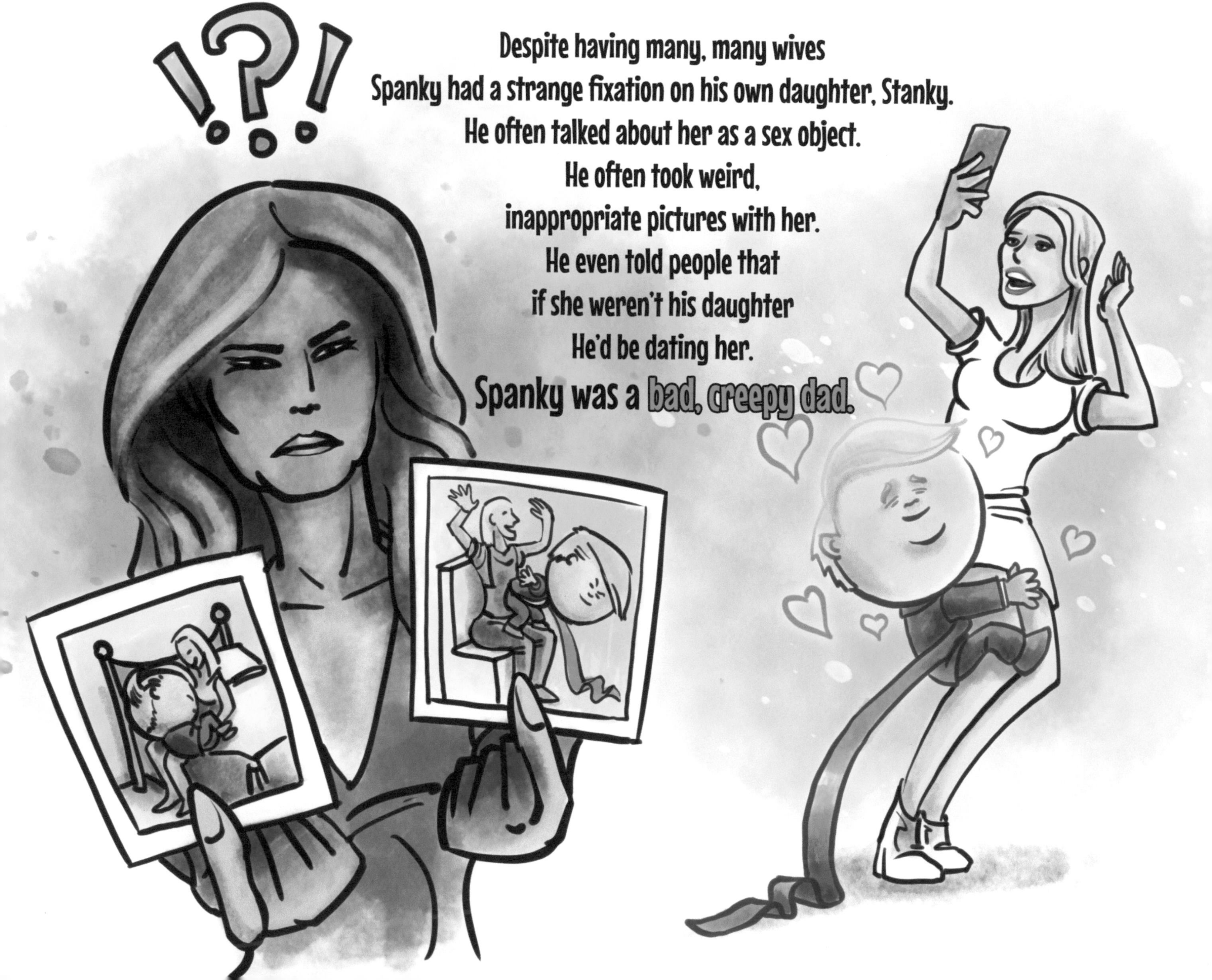

Spanky had a TV show called "The Appisstant."
People would have to prove they were dumb
enough to want to work for him.
If they weren't dumb enough
He pretended to fire them.
Some people watched, but not many.
It never won any awards,
and it was cancelled.

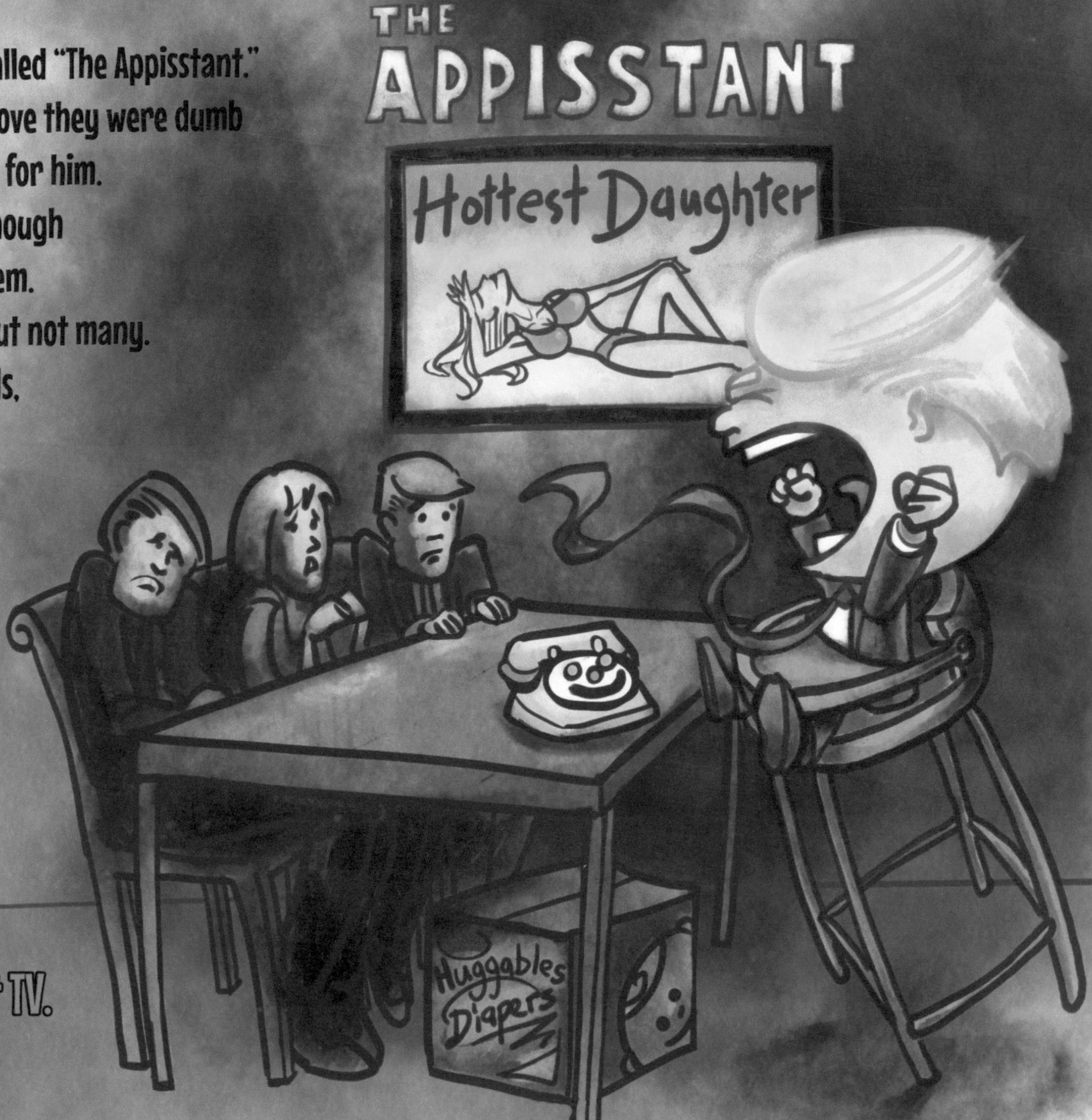
THE APPISSTANT
Hottest Daughter
Huggables Diapers

Spanky was **bad at TV.**

During that time, Spanky was on a bus.
He bragged to some guy about how he used to chase married women.
He bragged about sexual assault.
He bragged about how he would grab women's vaginas, and they would let him.
Spanky was being recorded the entire time.

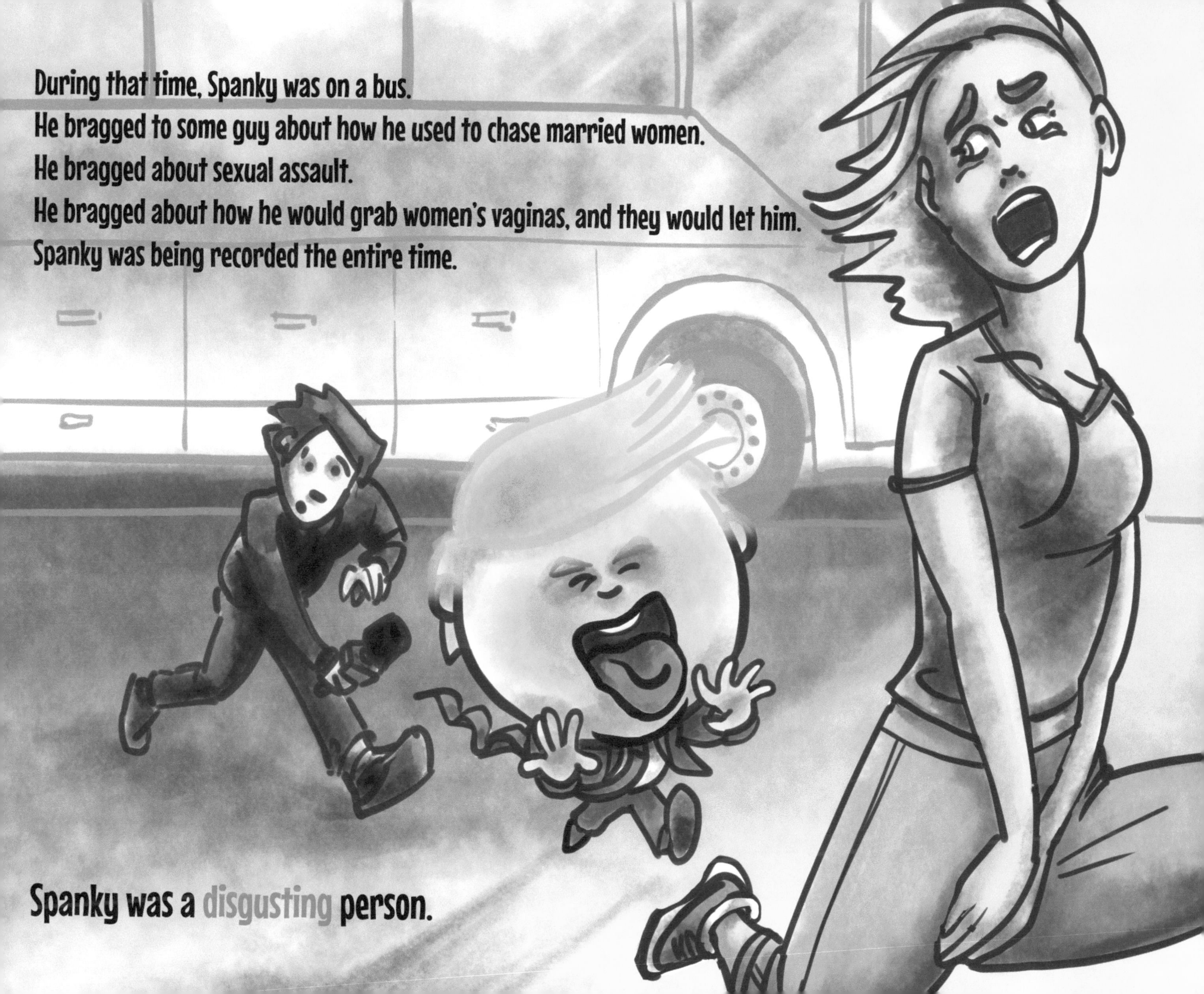

Spanky was a disgusting person.

One day, America elected a Great Man as President.
His name was Ourrock Nodrama.
He was intelligent, charismatic, dignified, handsome,
Honorable, an Inspirational Speaker, and a Family Man.
And he was America's FIRST Black President.
He was everything that Spanky wasn't.
Many people loved and admired him.

Spanky HATED him.

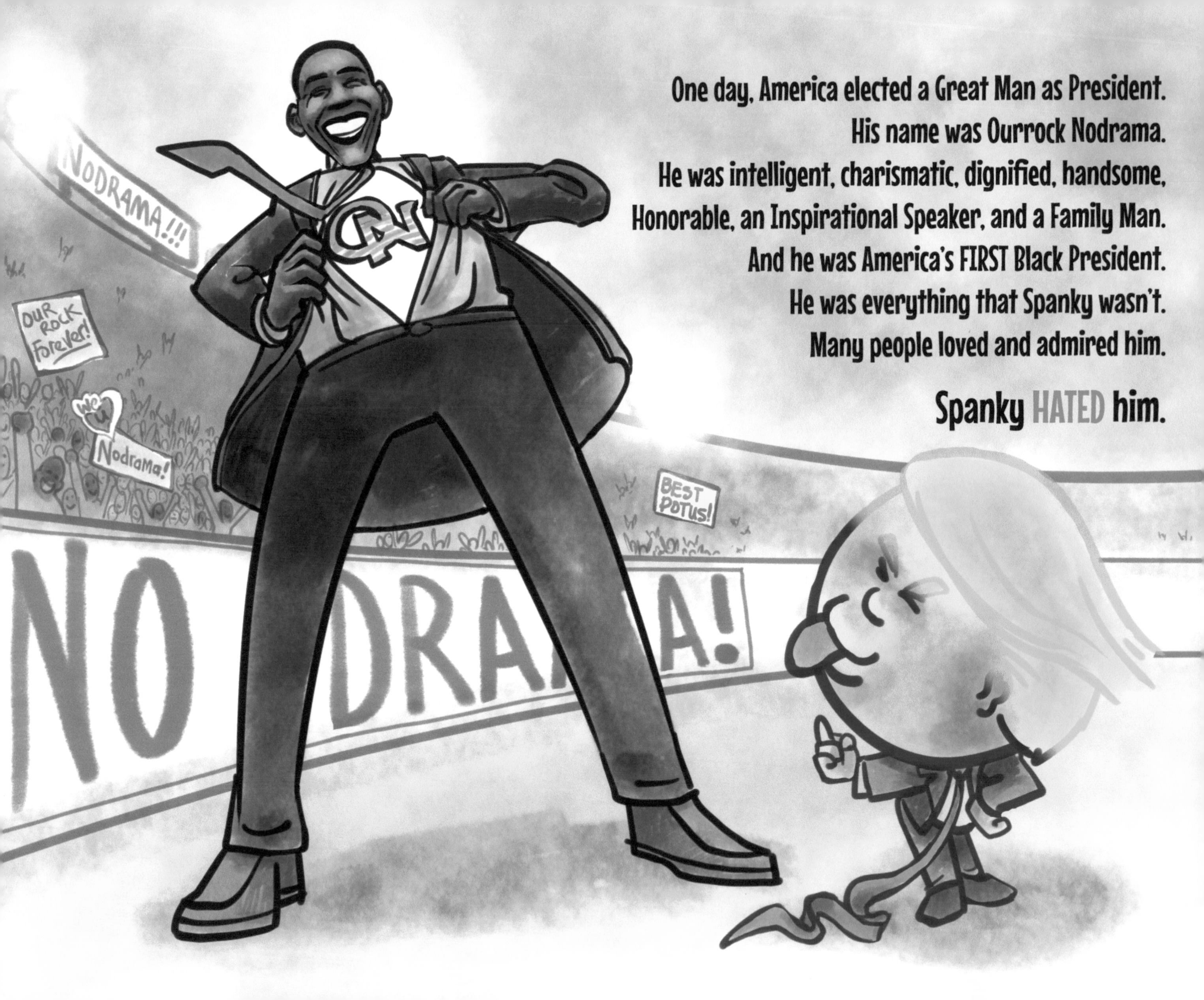

NODRAMA!!!
OUR ROCK Forever!
Nodrama!
BEST POTUS!

Despite the overwhelming optimism about President Nodrama,
there were a lot of racists in America who HATED the idea of a Black President. Spanky was one of them.
He made up a lie, saying that President Nodrama wasn't American,
and demanded to see his Birth Certificate.
And then, even after the President showed his Birth Certificate,
Spanky STILL insisted he wasn't American.

Spanky was a despicable person.

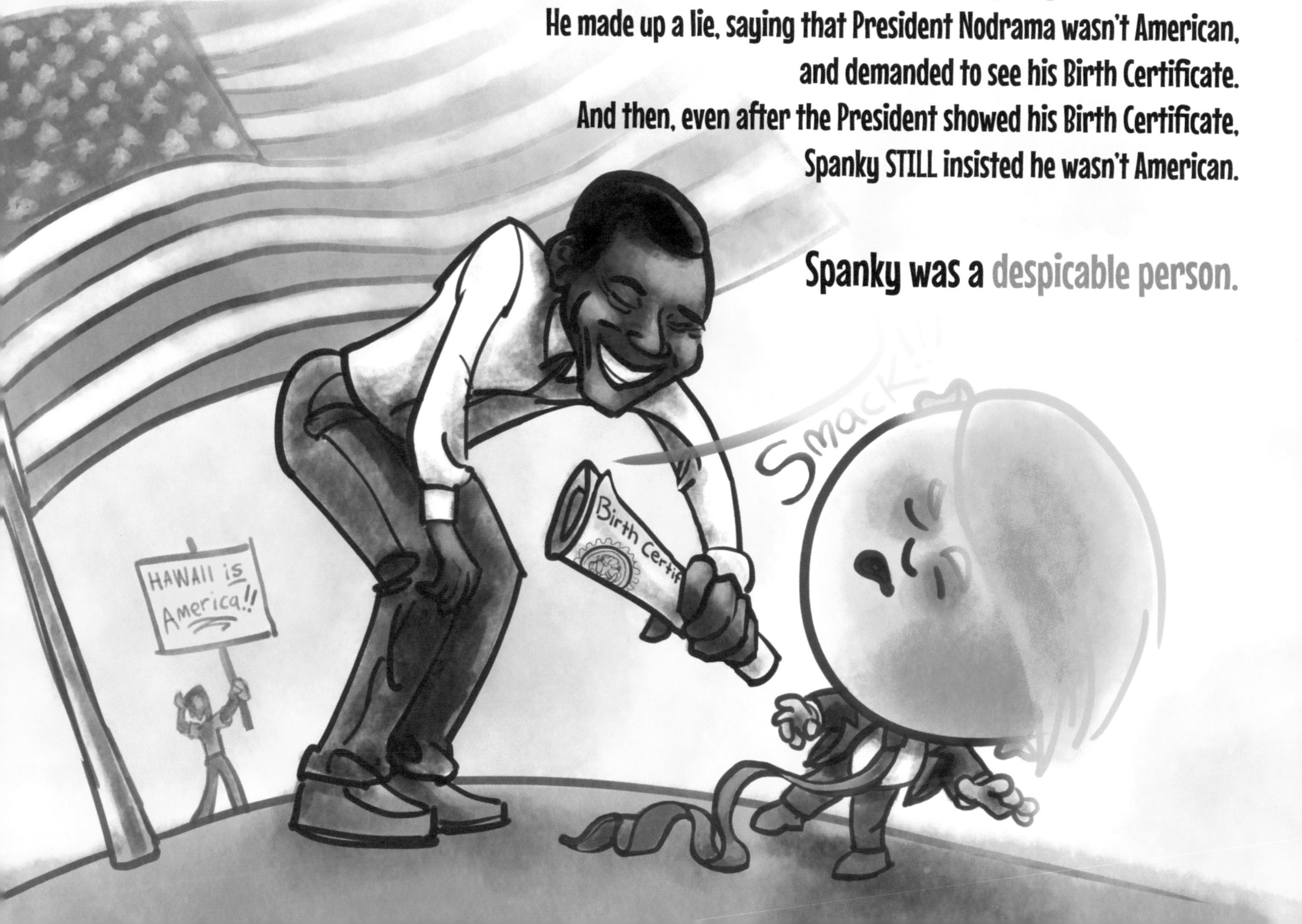
HAWAII is America!!
Birth Certif
Smack!!

Then, Spanky decided that he wanted to become President. He decided that there were just enough hateful people in America to follow him.

And when he announced he was running for President,

He told people he would build a big, beautiful wall to keep Mexicans out of America – and that Mexico would pay for it. He said they were sending murderers, rapists, and other criminals.

And horrible, hateful people cheered.

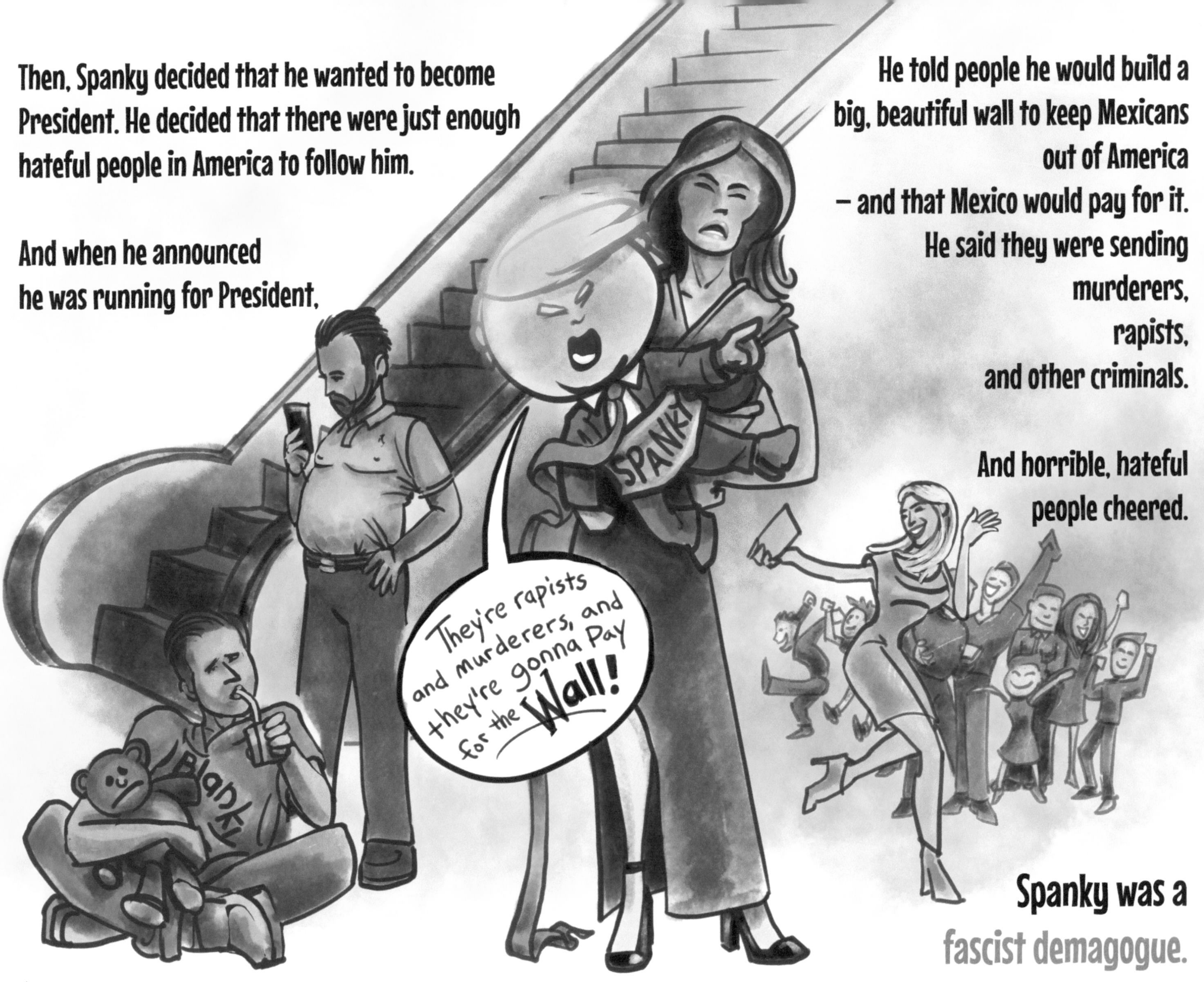
They're rapists and murderers, and they're gonna Pay for the Wall!
SPANKY
Blanky

Spanky was a **fascist demagogue.**

So Spanky went around the country saying the nastiest things about a great many people. He demeaned Senator John McCain, saying he wasn't a hero because he was captured.

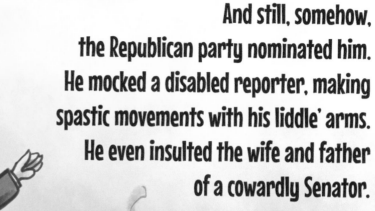

And still, somehow, the Republican party nominated him. He mocked a disabled reporter, making spastic movements with his liddle' arms. He even insulted the wife and father of a cowardly Senator.

Spanky was horrible, but his supporters were even worse.

Meanwhile, the Democrats nominated Emaillory Hinton for President.

She was a lawyer.
She was a former First Lady.
She was a US Senator.
She was the Secretary of State.

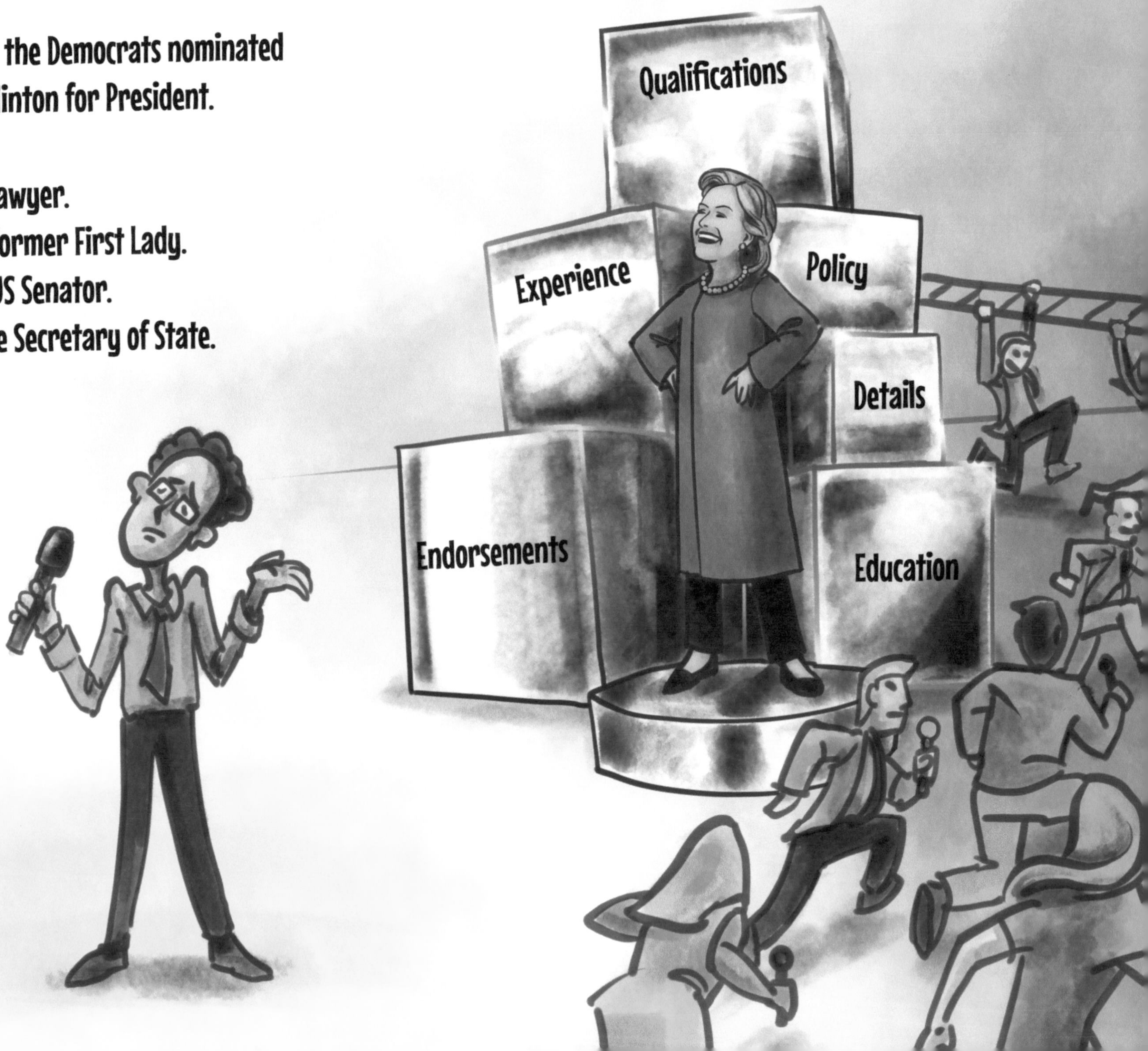
Qualifications
Experience
Policy
Details
Endorsements
Education

She was the MOST qualified candidate to ever run for President.
But Republicans made up a phony email controversy, and the media focused on that instead of anything else.

The media were REALLY helping to elect Spanky.

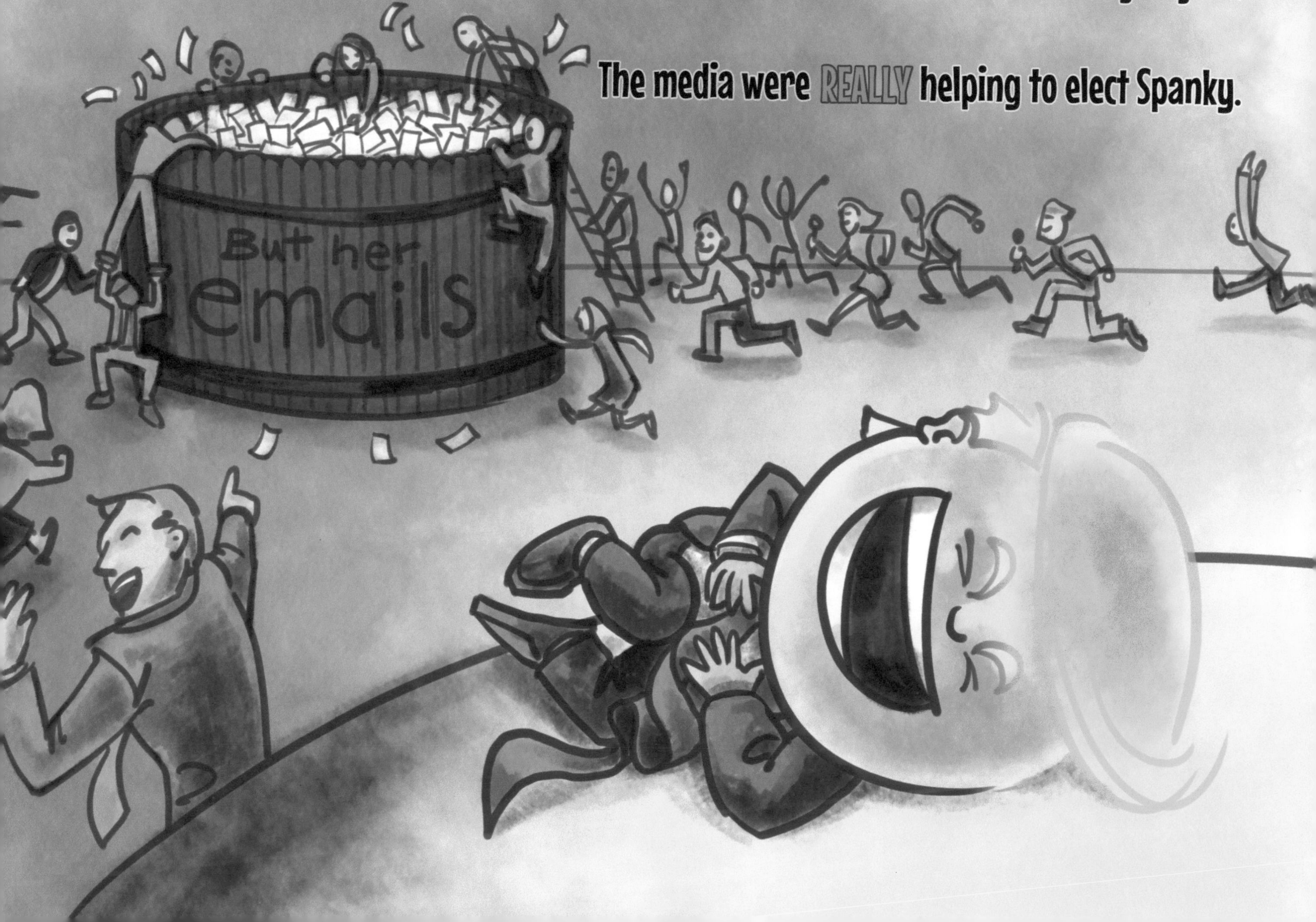
But her emails

Leaders around the world were watching the 2016 election with great interest. But perhaps nobody watched with more interest than Russia's President, Bladderbear Poopin. Poopin did not like Emaillory one bit. She was tough. Poopin didn't care much for Spanky, either, but he knew it would be MUCH easier to manipulate him. He'd started YEARS earlier, planting seeds of treason in Spanky's Liddle' brain. His manipulation was so complete, Spanky didn't even know he was being used.

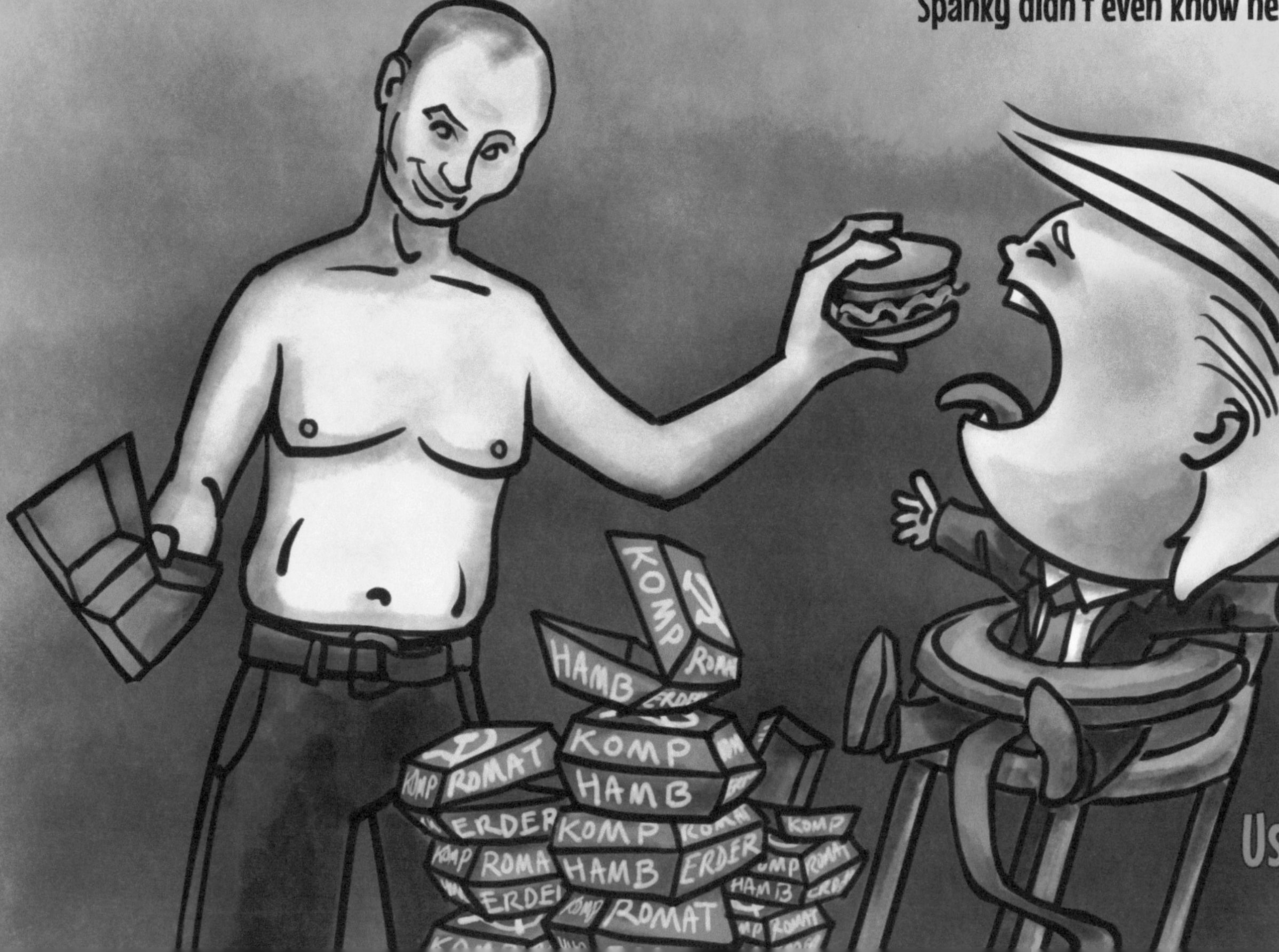
KOMP
ROMAT
HAMB
KOMP
HAMB
ERDER
KOMP
ROMAT

Spanky was a Useful Idiot.

In fact, Spanky was so blatantly compromised by Poopin, the FBI began investigating him during the presidential campaign. And then, during a campaign event, Spanky said to the camera, out of nowhere, "Russia, if you're listening, I hope you're able to find the 30,000 emails that are missing. I think you will probably be rewarded mightily by our press."

Spanky was asking Russia, a foreign enemy, to hack his opponent's emails.

Spanky was a traitor.

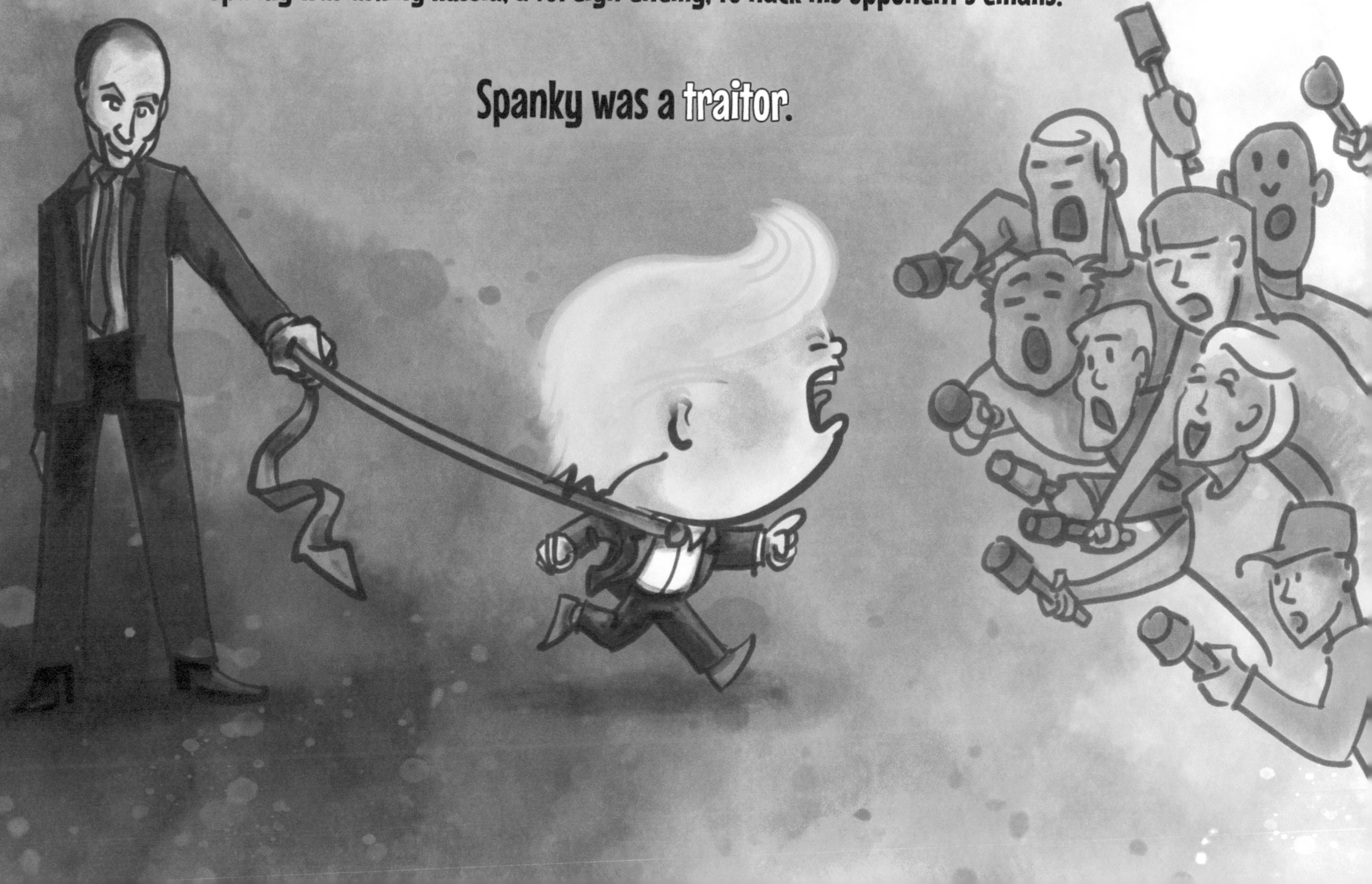

During the debates, Emaillory ran circles around Spanky. She was smart, she was funny, she was truthful. Spanky was dumb, he was sniffling, he was lying. And when Emaillory accused Spanky of being a puppet, He went ballistic, insisting he was "No puppet. No puppet!"

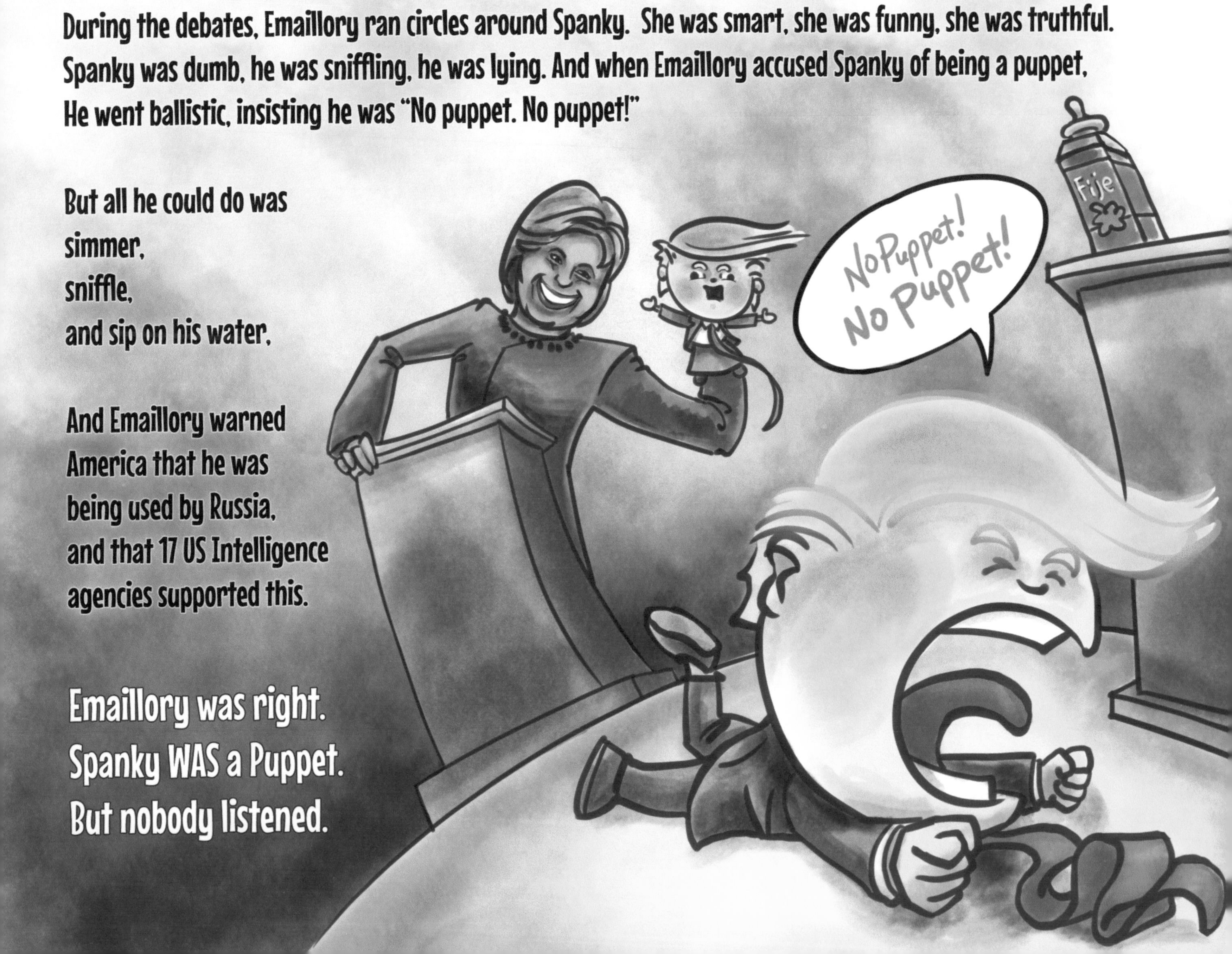
Fije
No Puppet!
No Puppet!

But all he could do was
simmer,
sniffle,
and sip on his water.

And Emaillory warned America that he was being used by Russia, and that 17 US Intelligence agencies supported this.

Emaillory was right.
Spanky WAS a Puppet.
But nobody listened.

A few months later, a videotape was found showing how Spanky bragged about grabbing Women's vaginas.
Everyone was SHOCKED. Everyone thought that was the end of Spanky's run.
But then WeakLeaks released some emails that fooled everyone,
And took the focus off of Spanky...again.
And the race was closer than ever.

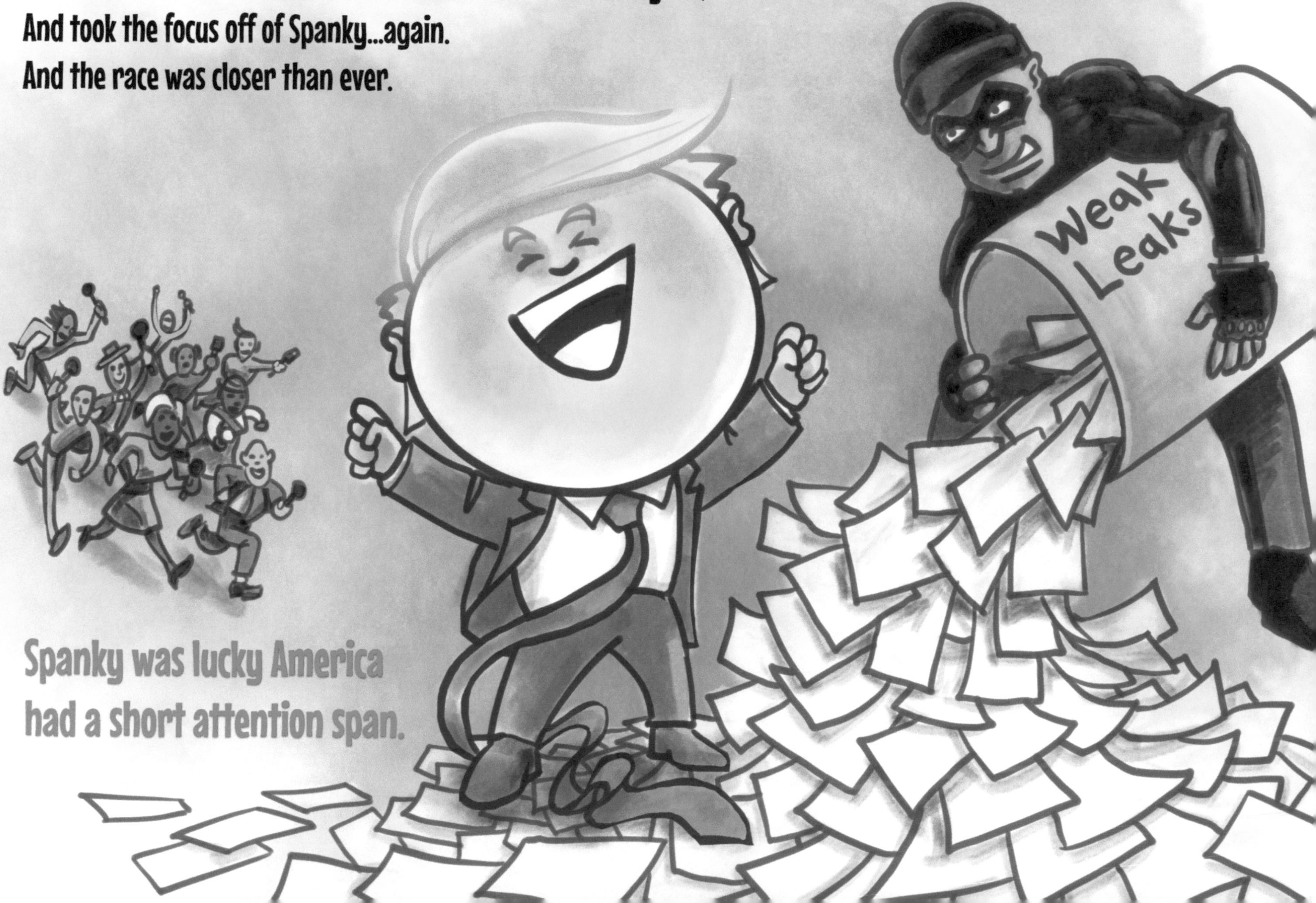
Weak Leaks

Spanky was lucky America
had a short attention span.

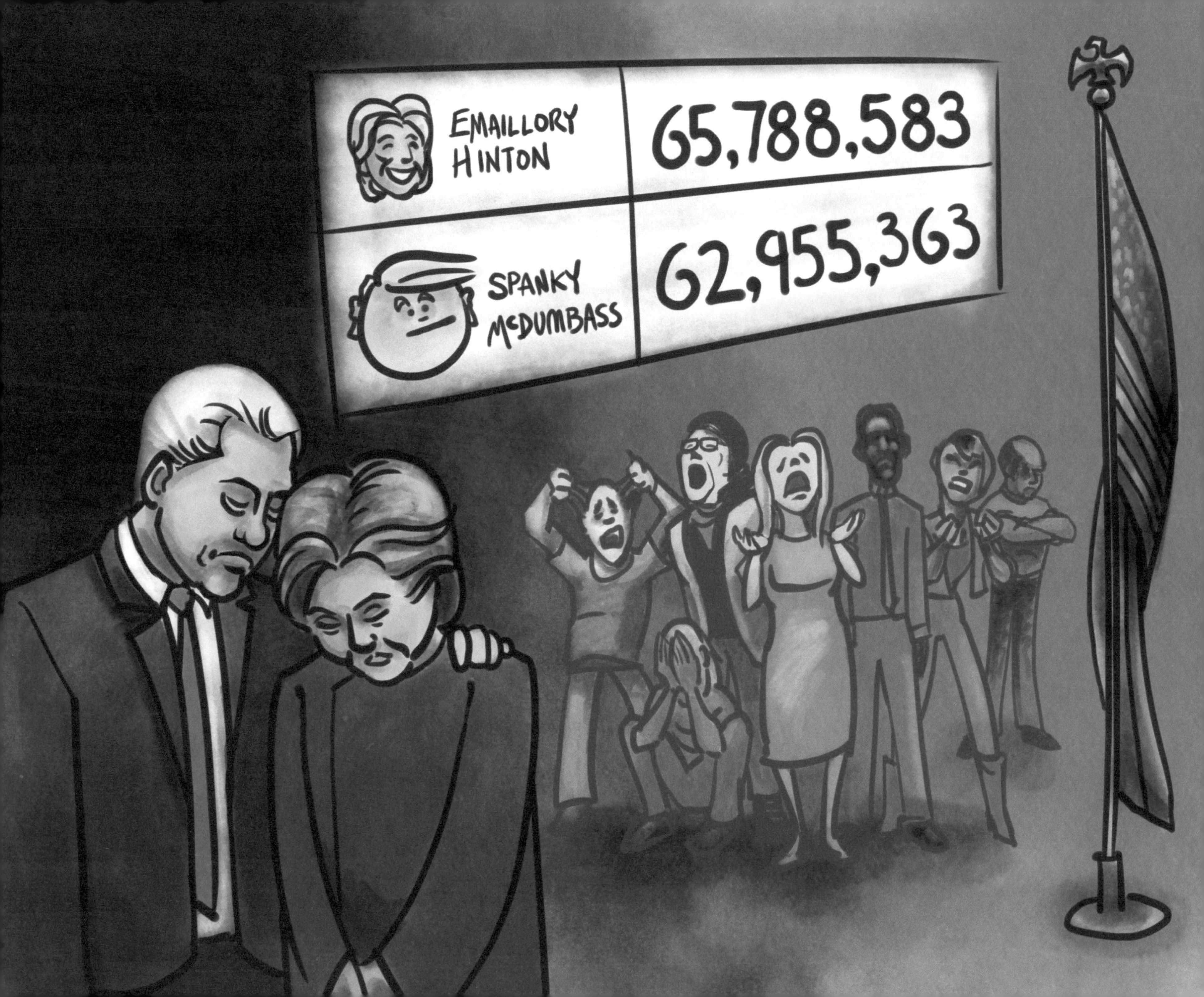
EMAILLORY HINTON
65,788,583
SPANKY McDUMBASS
62,955,363

When Election Day finally came,
Emaillory WON the Popular Vote
by 3 million people!

But she lost the Electoral Vote,
which is an old system
that was created to make sure
slave owners had a "fair shake."

Spanky was now...

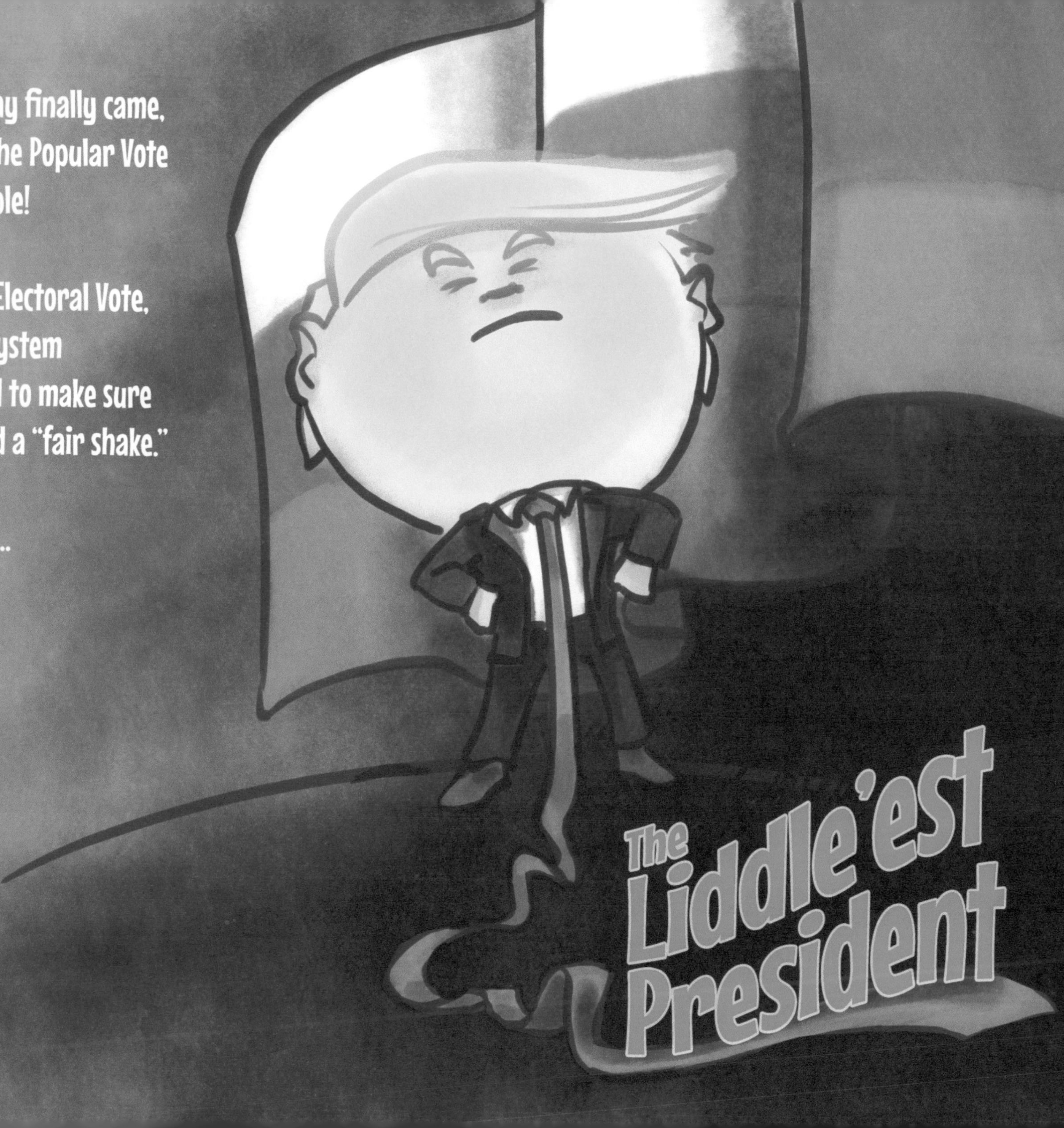
The Liddle'est President

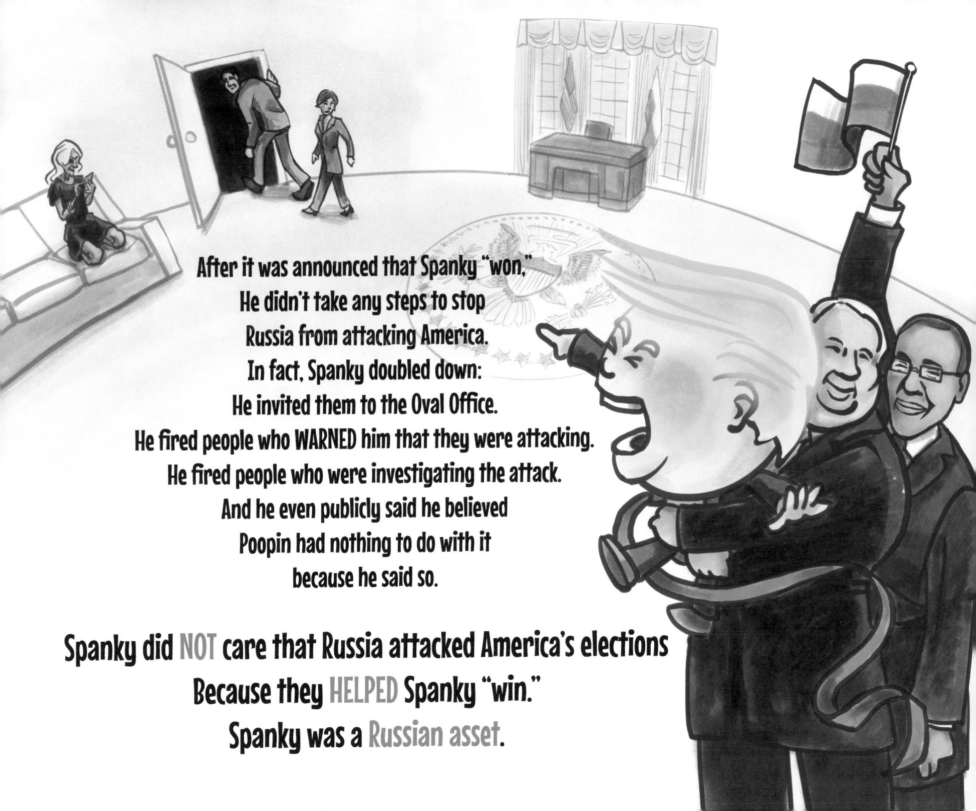

After it was announced that Spanky "won,"
He didn't take any steps to stop
Russia from attacking America.
In fact, Spanky doubled down:
He invited them to the Oval Office.
He fired people who WARNED him that they were attacking.
He fired people who were investigating the attack.
And he even publicly said he believed
Poopin had nothing to do with it
because he said so.

Spanky did NOT care that Russia attacked America's elections
Because they HELPED Spanky "win."
Spanky was a Russian asset.

For the next 3 years, Spanky did not spend any time trying to win over voters who DIDN'T vote for him.

Instead, Spanky spent his time:
Trying to steal Americans' healthcare
Giving the rich MASSIVE tax breaks
Starting ridiculous trade wars
Locking babies in cages
Violating emoluments
Obstructing justice
Coddling Nazis
Inciting violence
Insulting allies
Golfing

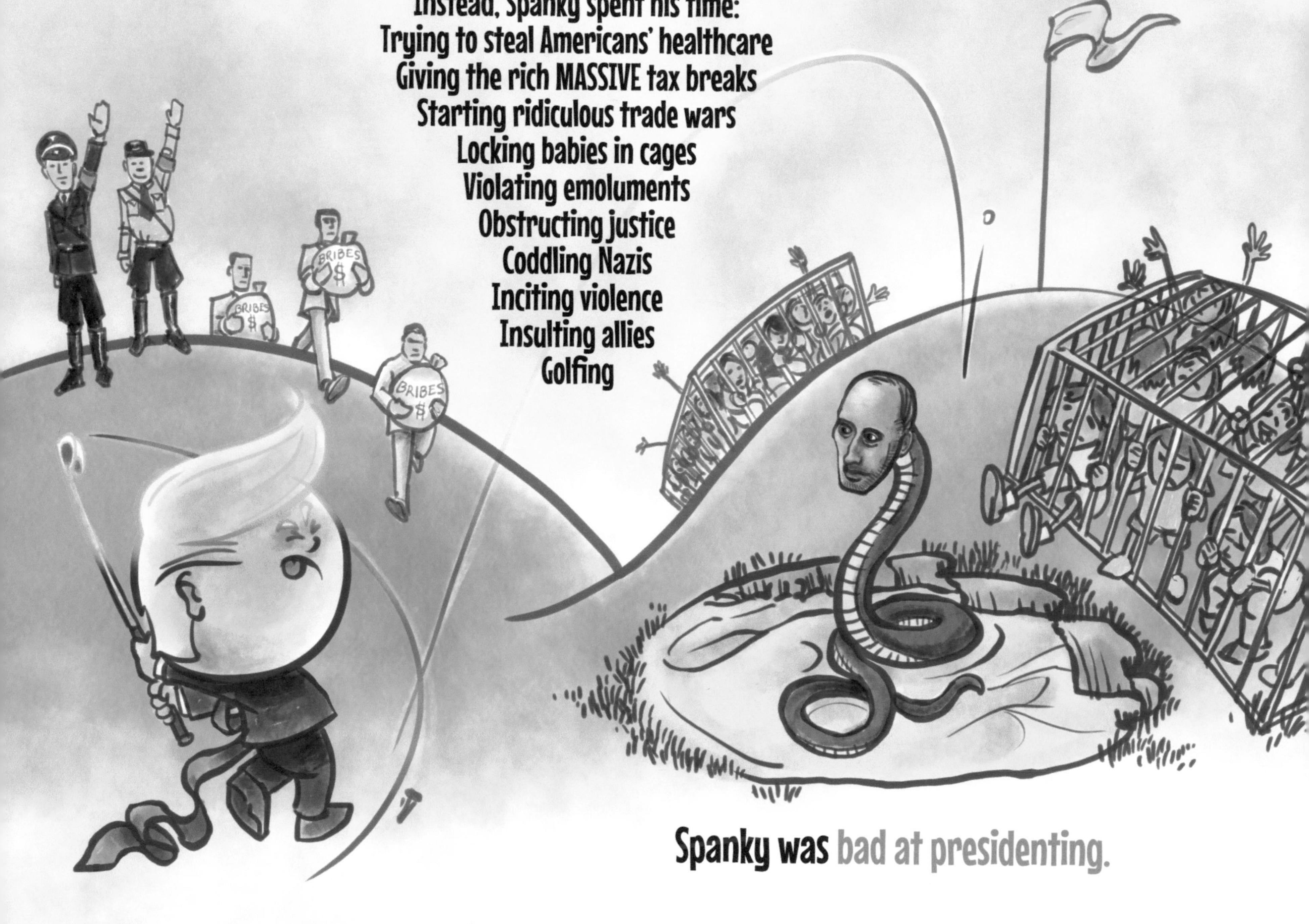

BRIBES $
BRIBES $
BRIBES $

Spanky was bad at presidenting.

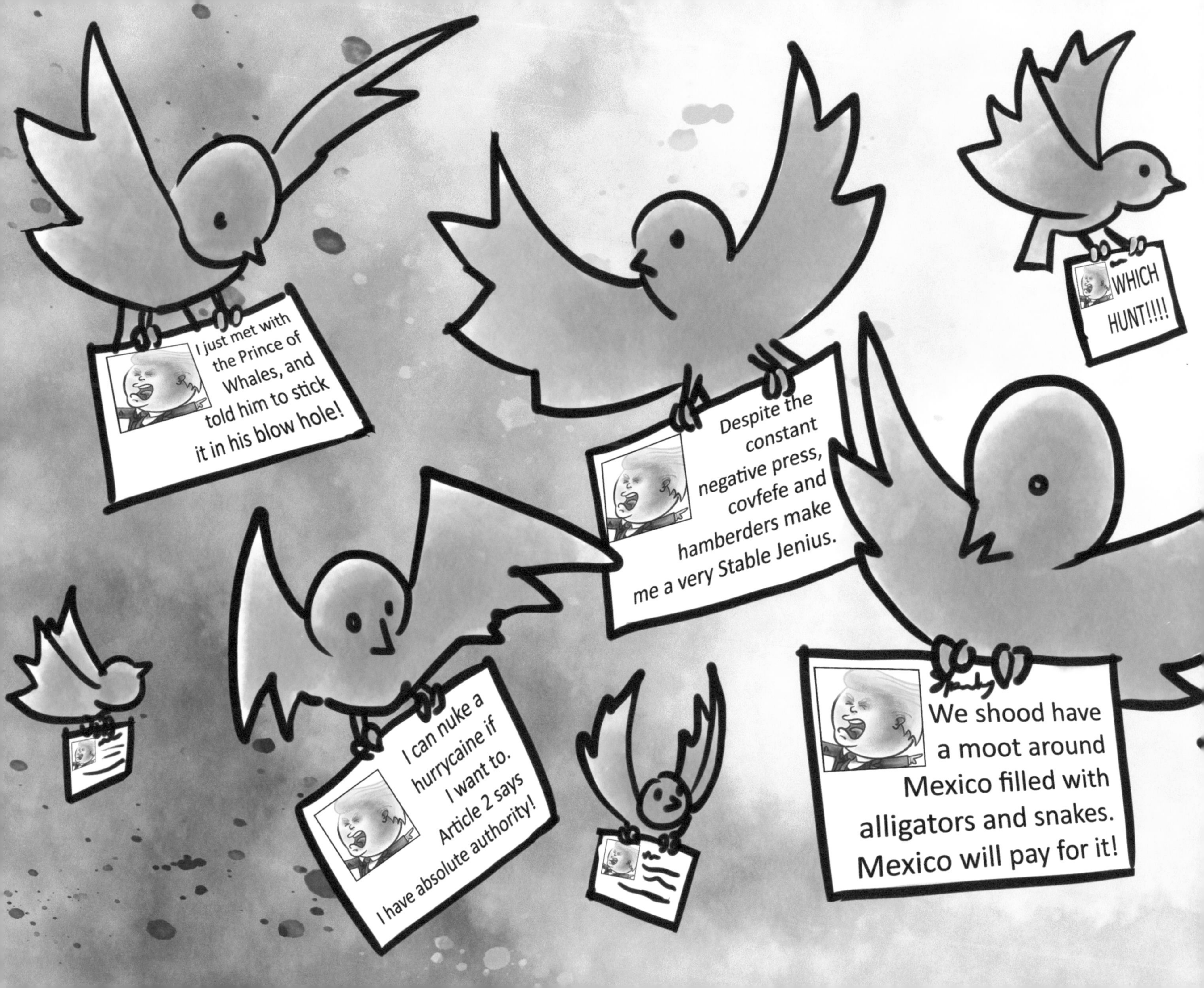
I just met with the Prince of Whales, and told him to stick it in his blow hole!
Despite the constant negative press, covfefe and hamberders make me a very Stable Jenius.
WHICH HUNT!!!!
I can nuke a hurrycaine if I want to. Article 2 says I have absolute authority!
We shood have a moot around Mexico filled with alligators and snakes. Mexico will pay for it!

But what Spanky loved doing, more than anything, was TWEETING.
Many of his tweets were complaints, like "WITCH HUNT!"
or "PRESIDENTIAL HARASSMENT!"
Other tweets complained about celebrities,
or Fake News, or the size of his crowds.
Sometimes he attacked people for no reason.
Sometimes he threatened people.
Once, he even threatened Civil War.

Spanky was bad at tweeting.

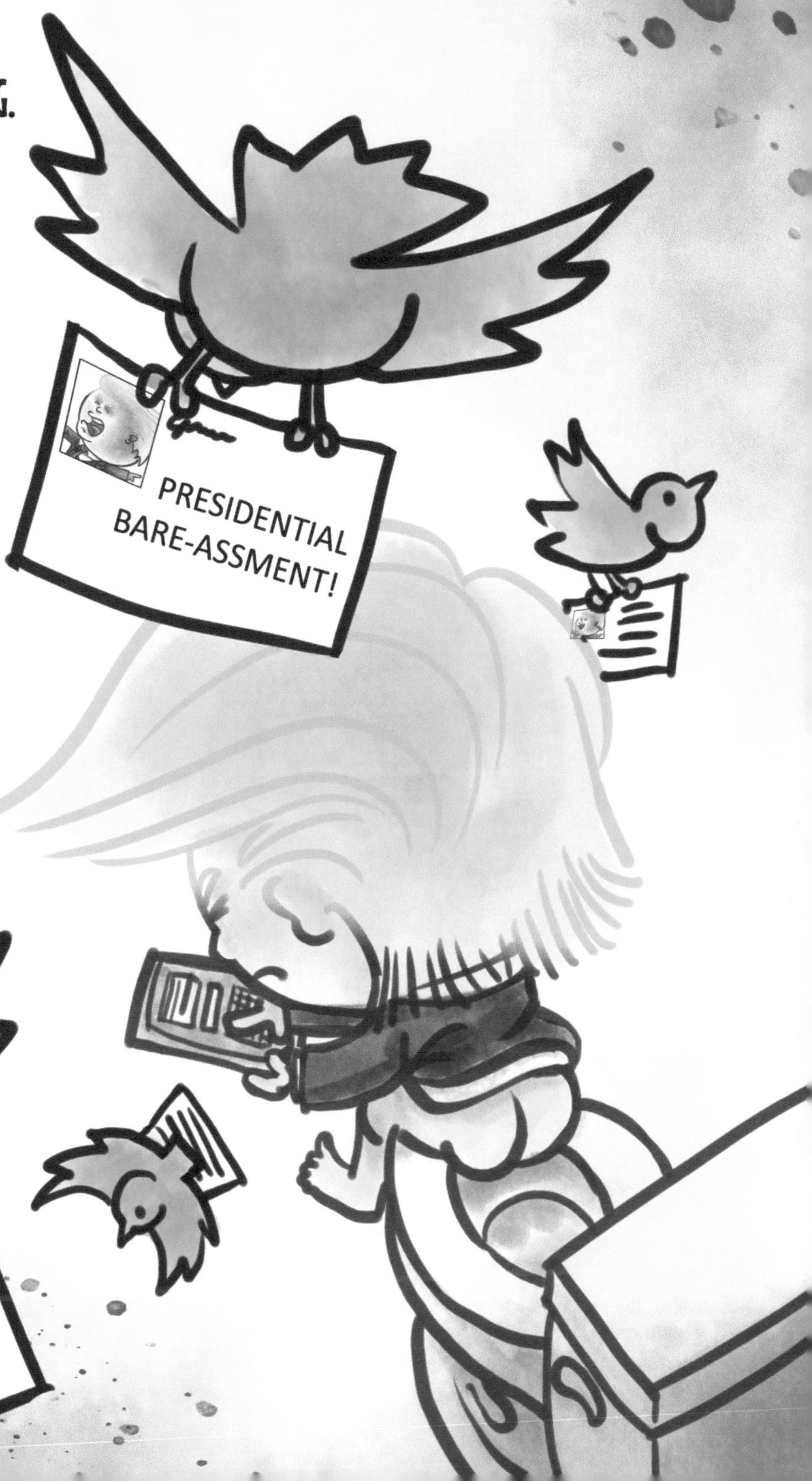
PRESIDENTIAL BARE-ASSMENT!

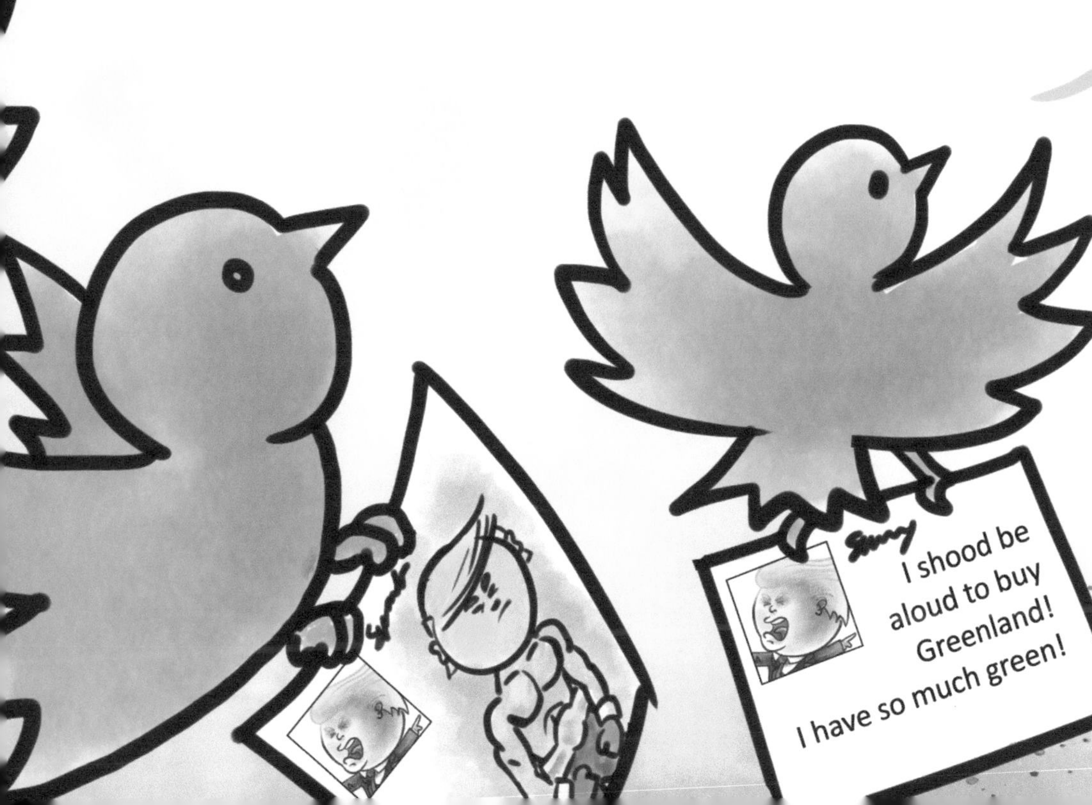
I shood be aloud to buy Greenland! Greenland!
I have so much green!

The only thing Spanky was "good" at was breaking the law. He just wasn't very good at hiding it. One day, when he heard that former VP Moe Reyden was running for President, Spanky got worried. REAL worried. He was so worried, he called the President of Ukraine and asked him to investigate Reyden in exchange for military aid that was already approved by Congress.

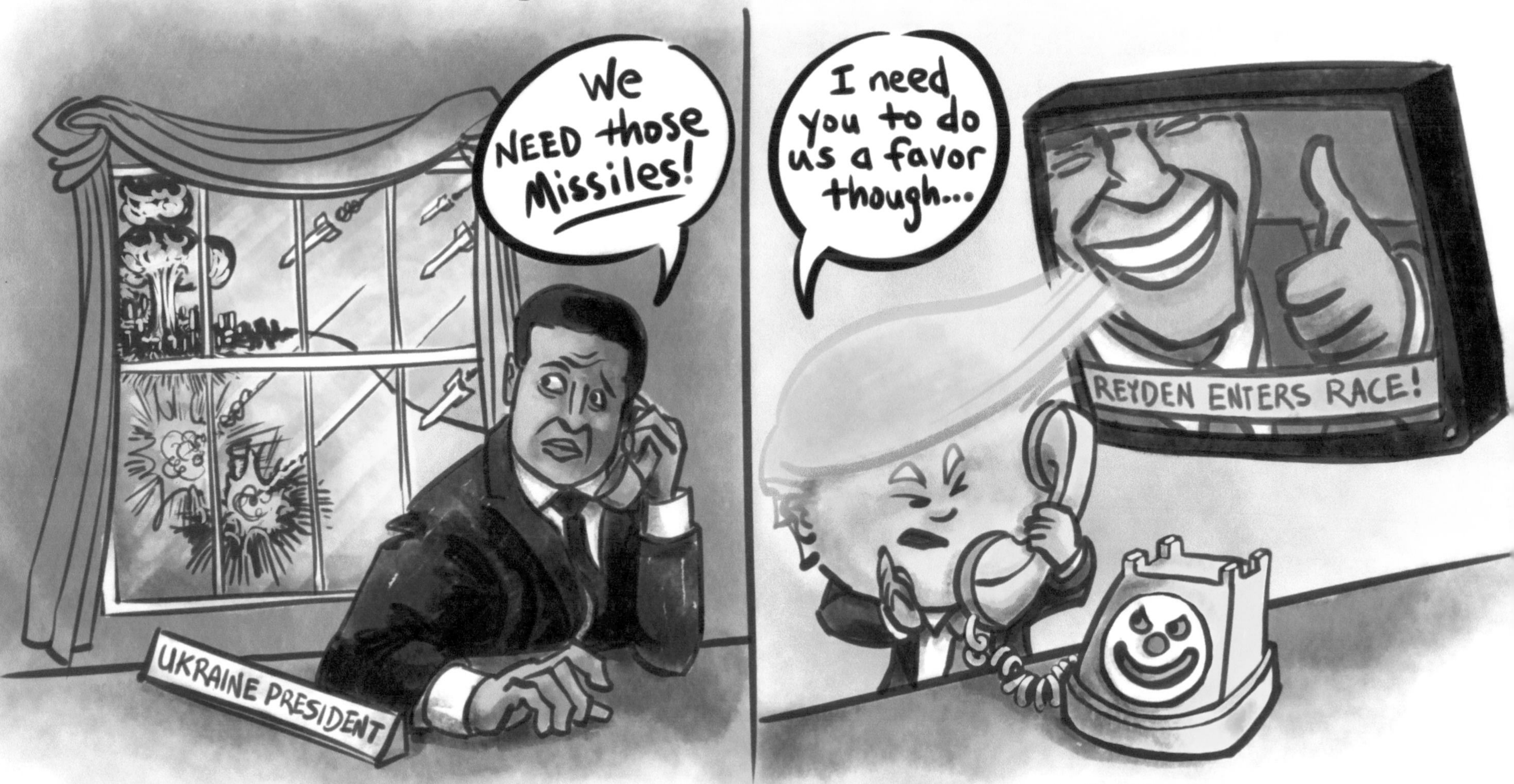

We NEED those Missiles!
I need you to do us a favor though...
REYDEN ENTERS RACE!
UKRAINE PRESIDENT

Asking a foreign country to investigate your political rival is illegal.
Even for a President. ESPECIALLY for a President.

Because nobody is above the law.

Spanky would have gotten away with it, but a Whistleblower stepped forward, and told people about Spanky's call. Spanky was mad, and said it was a "perfect call."
He finally released Ukraine's military aid, but only because he got caught.

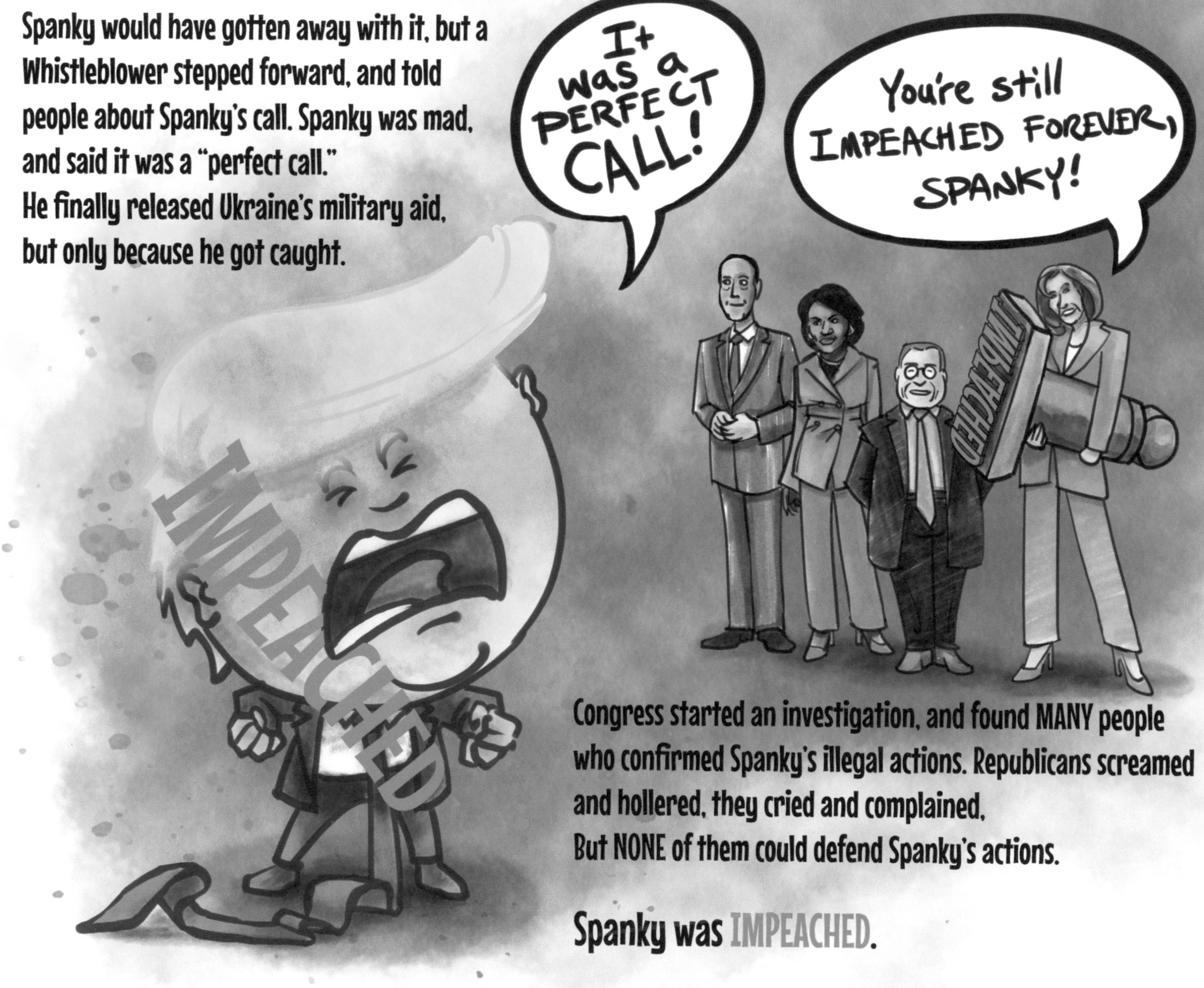

It was a PERFECT CALL!
You're still IMPEACHED FOREVER, SPANKY!
IMPEACHED
IMPEACHED

Congress started an investigation, and found MANY people who confirmed Spanky's illegal actions. Republicans screamed and hollered, they cried and complained,
But NONE of them could defend Spanky's actions.

Spanky was IMPEACHED.

But when Speaker Fancy Virtuosi finally sent the Articles of Impeachment to the Senate,
Senator Turtle McCoward turned the Senate trial into a farce.
He refused to let any witnesses testify,
And Spanky was "exonerated"
without really being exonerated.

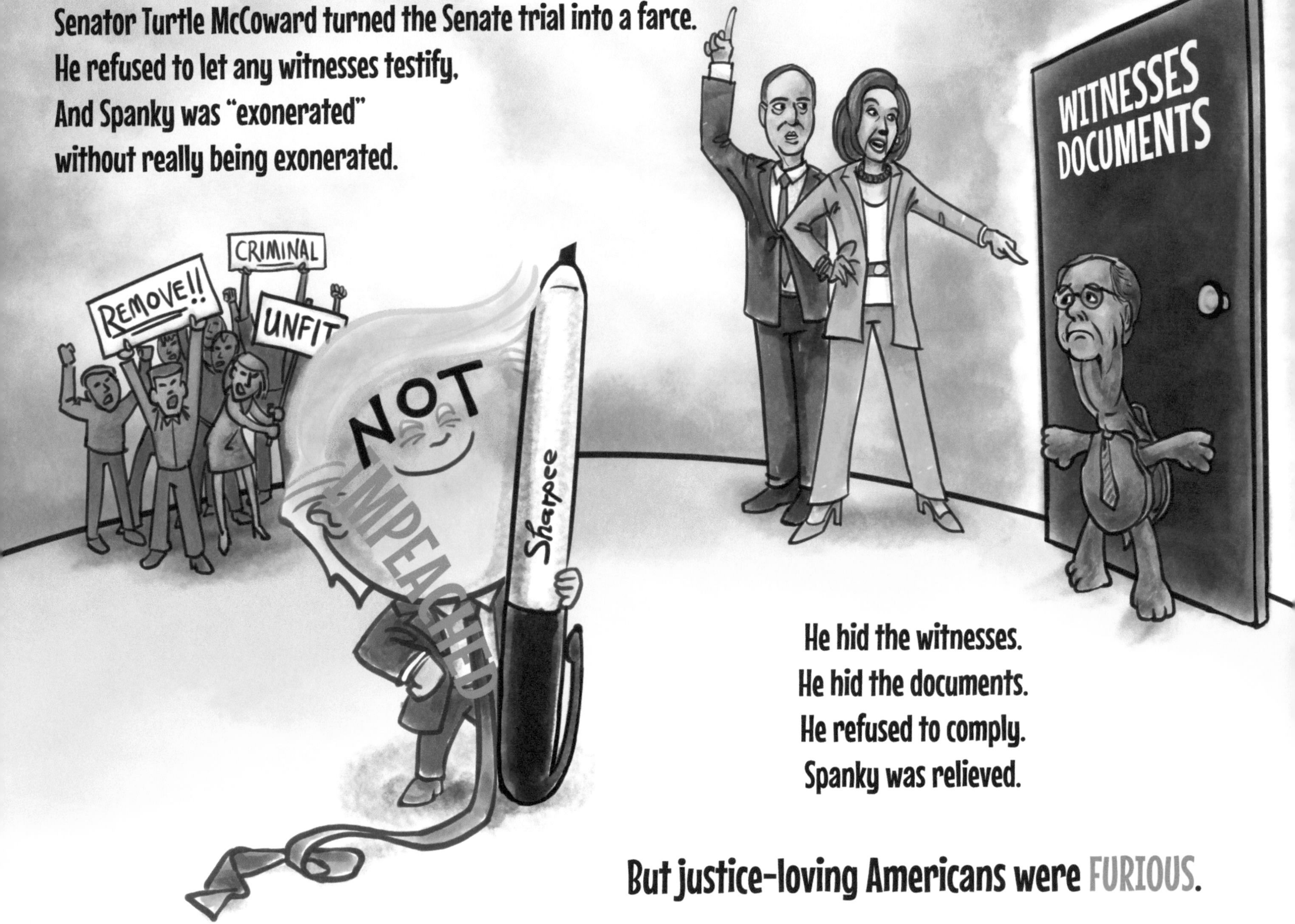

CRIMINAL
REMOVE!!
UNFIT
NOT
IMPEACHED
Sharpee
WITNESSES
DOCUMENTS

He hid the witnesses.
He hid the documents.
He refused to comply.
Spanky was relieved.

But justice-loving Americans were FURIOUS.

Democrats were so angry at the fake trial, they quickly fell in line behind ONE Presidential contender: Candidate Somebody Blue. This candidate wasn't everyone's first choice; many people wanted Someone Else. Some people wanted to vote for Anyone Else. But they knew that if Spanky won, he'd just keep on breaking the law and not being held accountable for it. So the only thing folks could agree upon was that Somebody Blue was the only person who could defeat Spanky. Even Republicans and Independents agreed on this.

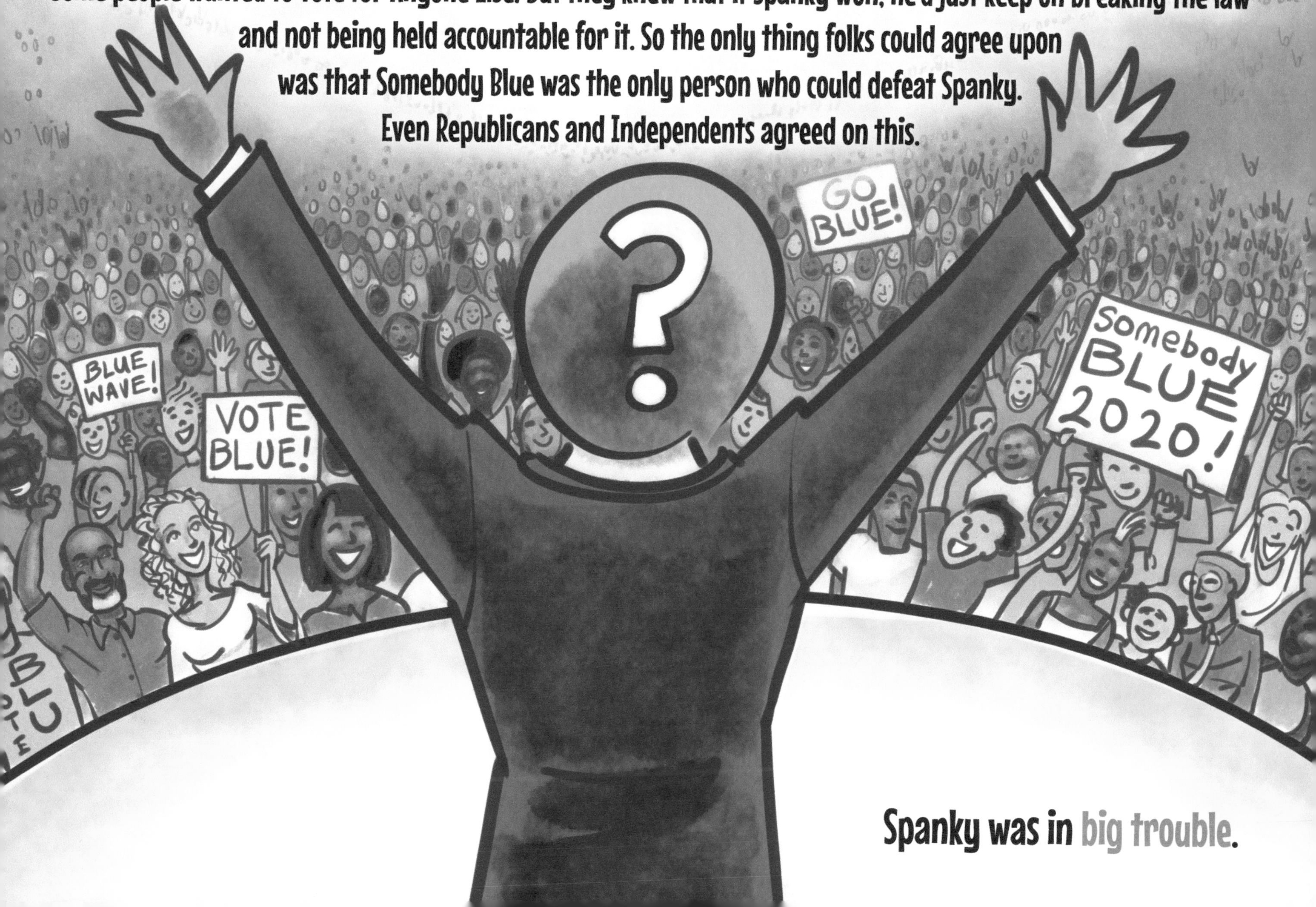
GO BLUE!
BLUE WAVE!
VOTE BLUE!
Somebody BLUE 2020!

Spanky was in big trouble.

Americans were so infuriated by the verdict of the trial, they quickly mobilized to beat Spanky in 2020.

Hundreds volunteered.

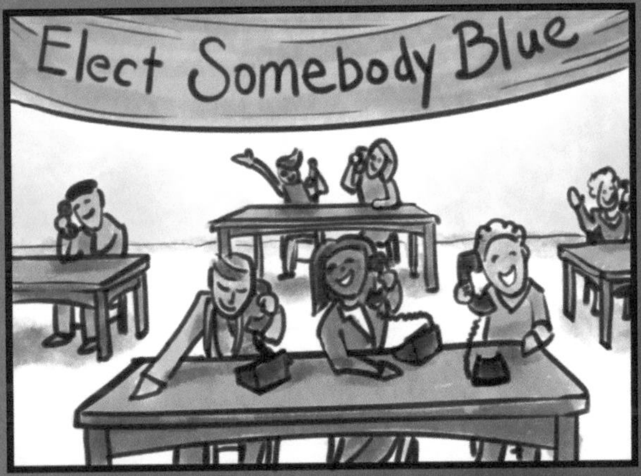
Elect Somebody Blue

Thousands registered to vote.

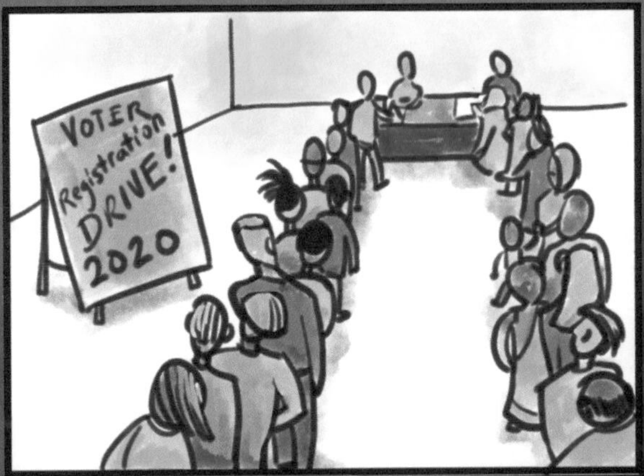
VOTER Registration DRIVE! 2020

Millions marched to spread the word.

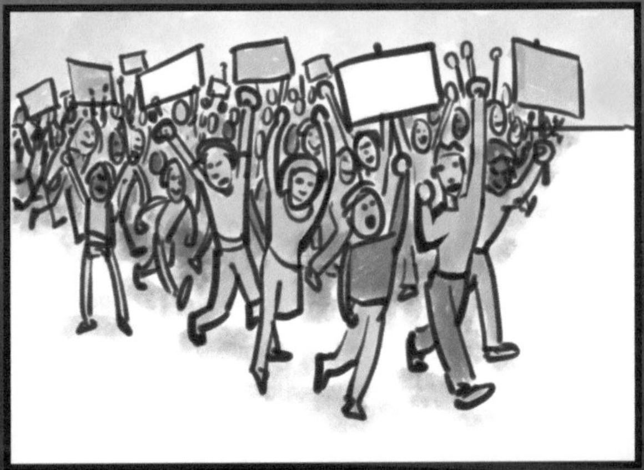

Spanky did everything he could to change people's minds.

He lied. He screamed.
He tweeted. He bleated.

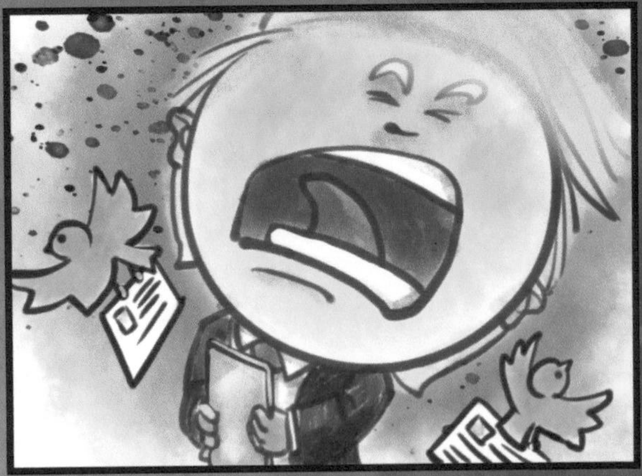

He even tried to start a war or two.

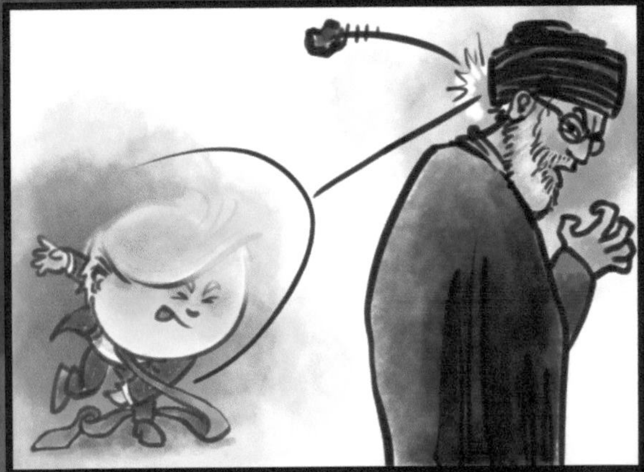

But America had already made up its mind.

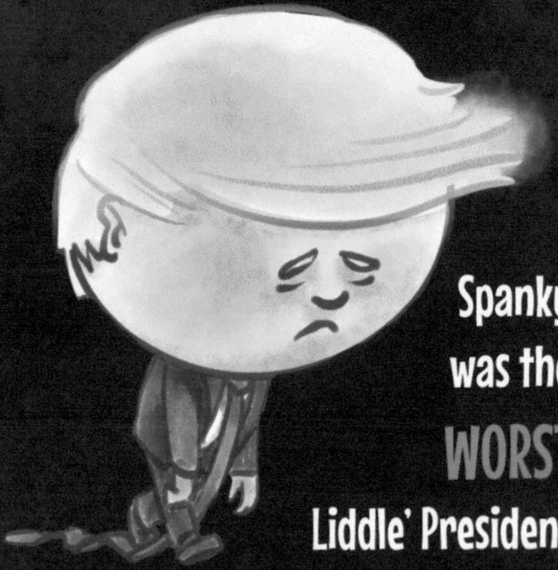

Spanky was the **WORST** Liddle' President America had ever seen.

And, on election night,
The state election results came in,
And Candidate Blue kept winning states.
And Spanky was worried.

VERY worried.

The results kept coming in,
and then the news stations announced
that the Liddle'est President had just
LOST in the Bigliest landslide in history.

The People had spoken.

Spanky was a LOSER.

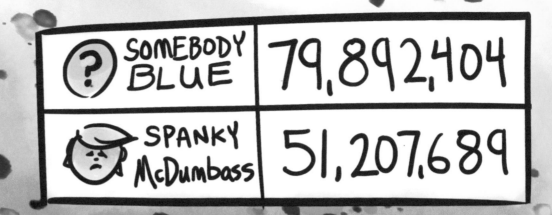

| ? SOMEBODY BLUE | 79,892,404 |
| SPANKY McDumbass | 51,207,689 |

On the day after election day,
The People let out a HUGE cheer and rejoiced.
Many stayed home from work.
They danced.
They sang.
They laughed.
They cried.

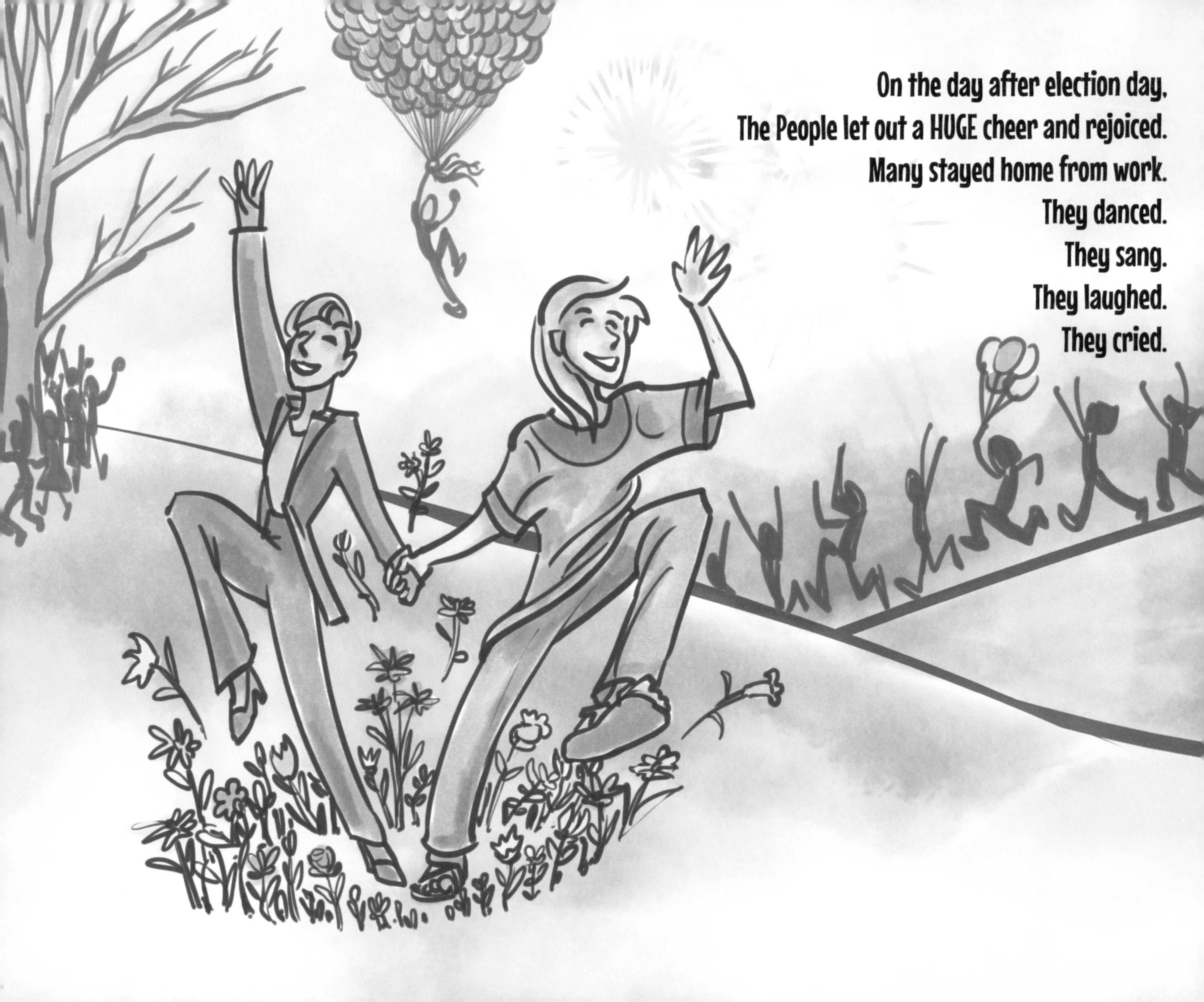

The biggest nightmare
America had ever
seen was over.
But Spanky's
nightmare
was just
beginning.

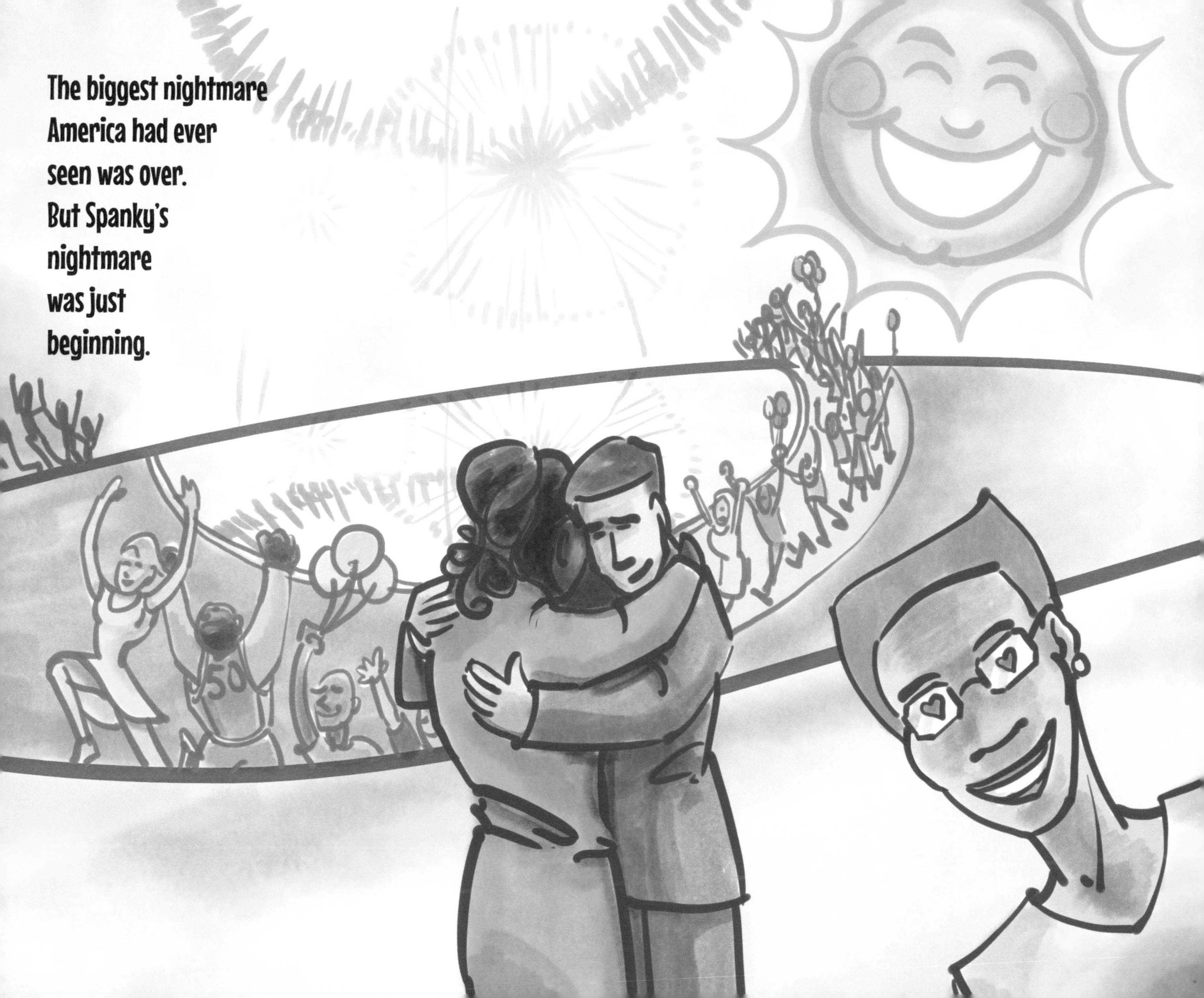
50

Spanky decided that he wasn't going to leave the White House.
He stomped his feet and yelled "Fake News! Presidential Harassment!"
He wouldn't accept the fact that he had lost and very possibly would have to go to jail.
He tried to say that the election was rigged, but he had lost by such a HUGE margin, there was no doubt.

Spanky would HAVE to leave.

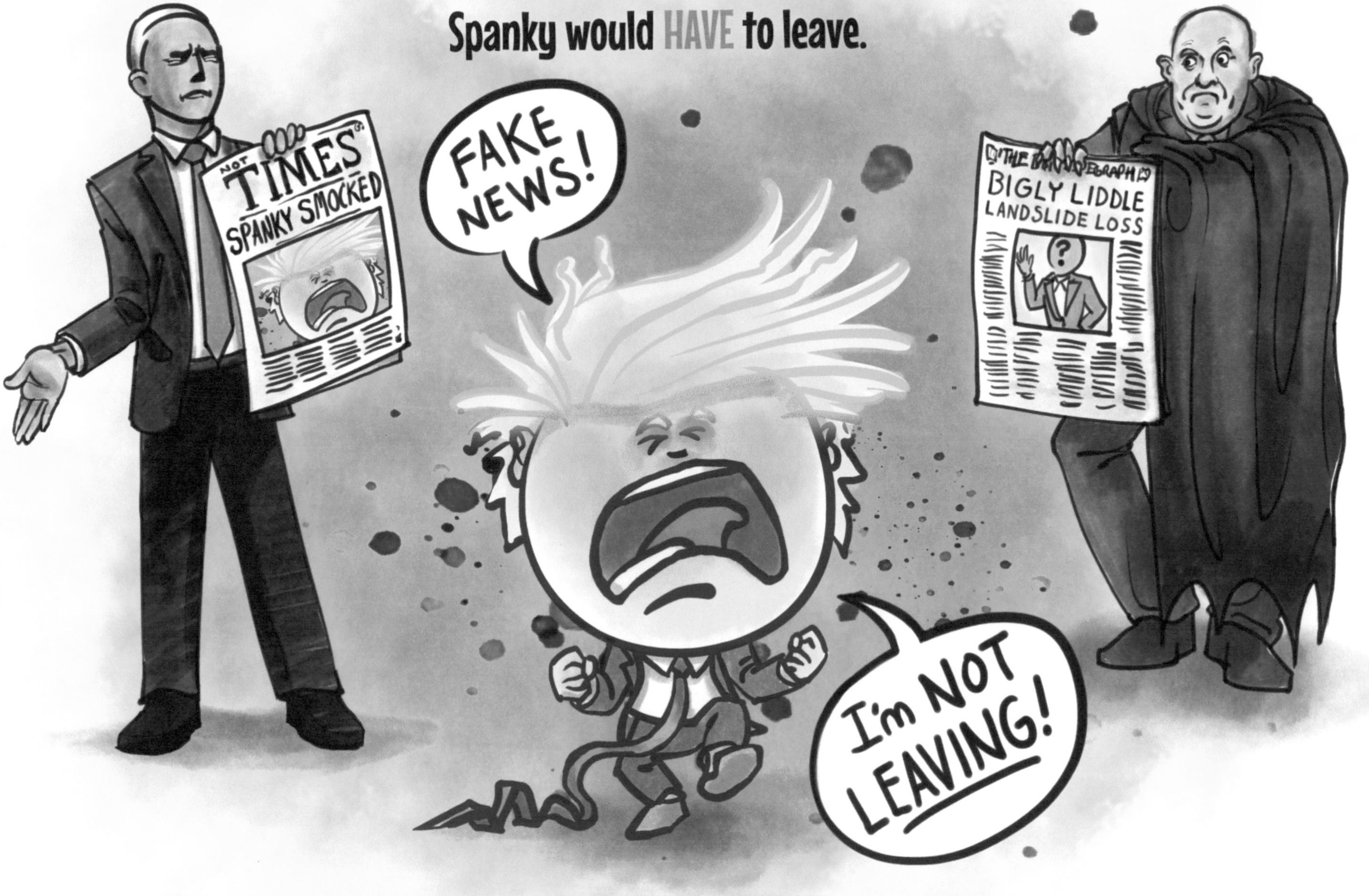
NOT TIMES
SPANKY SMOCKED
FAKE NEWS!
BIGLY LIDDLE LANDSLIDE LOSS
I'm NOT LEAVING!

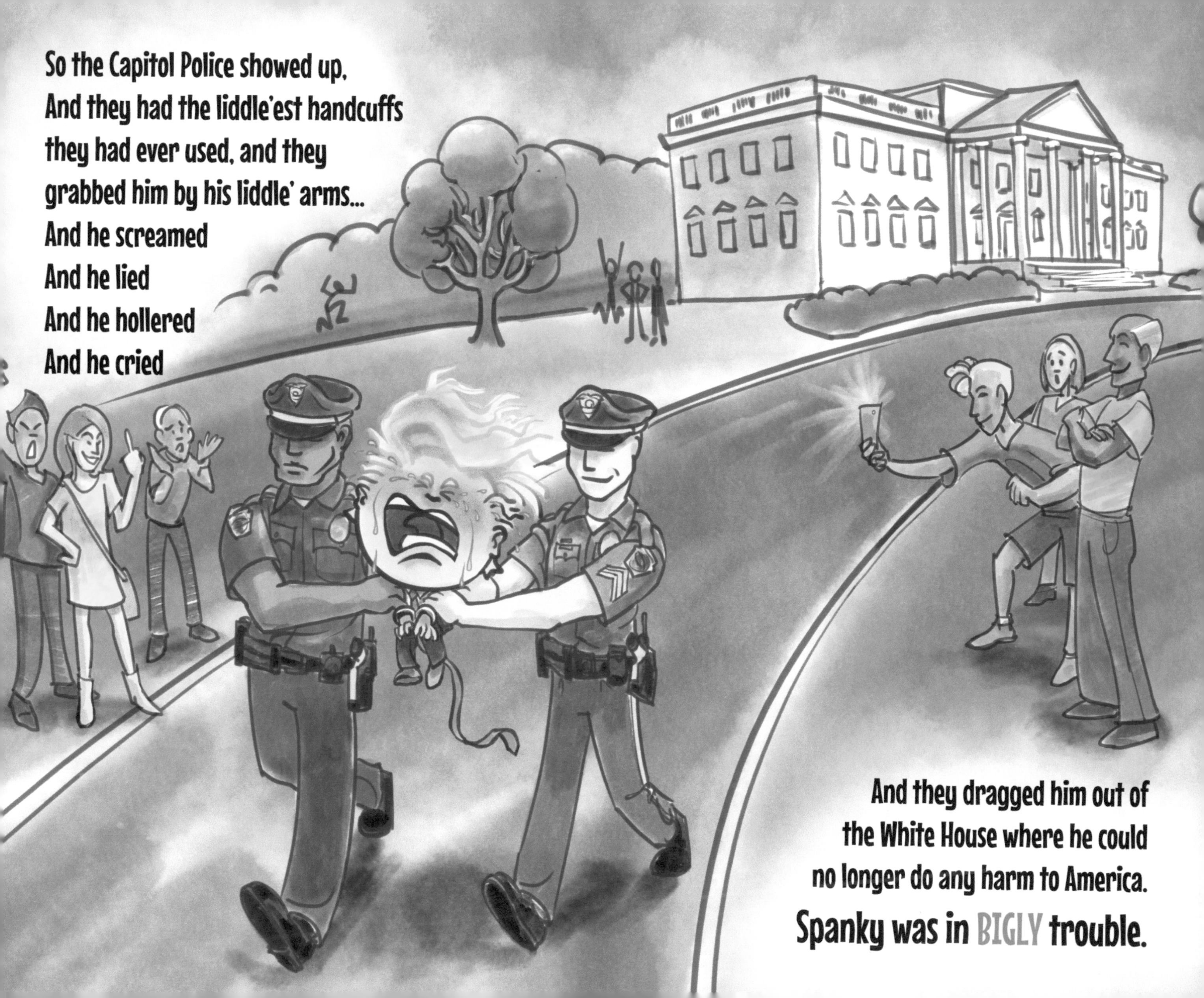

So the Capitol Police showed up,
And they had the liddle'est handcuffs
they had ever used, and they
grabbed him by his liddle' arms...
And he screamed
And he lied
And he hollered
And he cried

And they dragged him out of
the White House where he could
no longer do any harm to America.

Spanky was in BIGLY trouble.

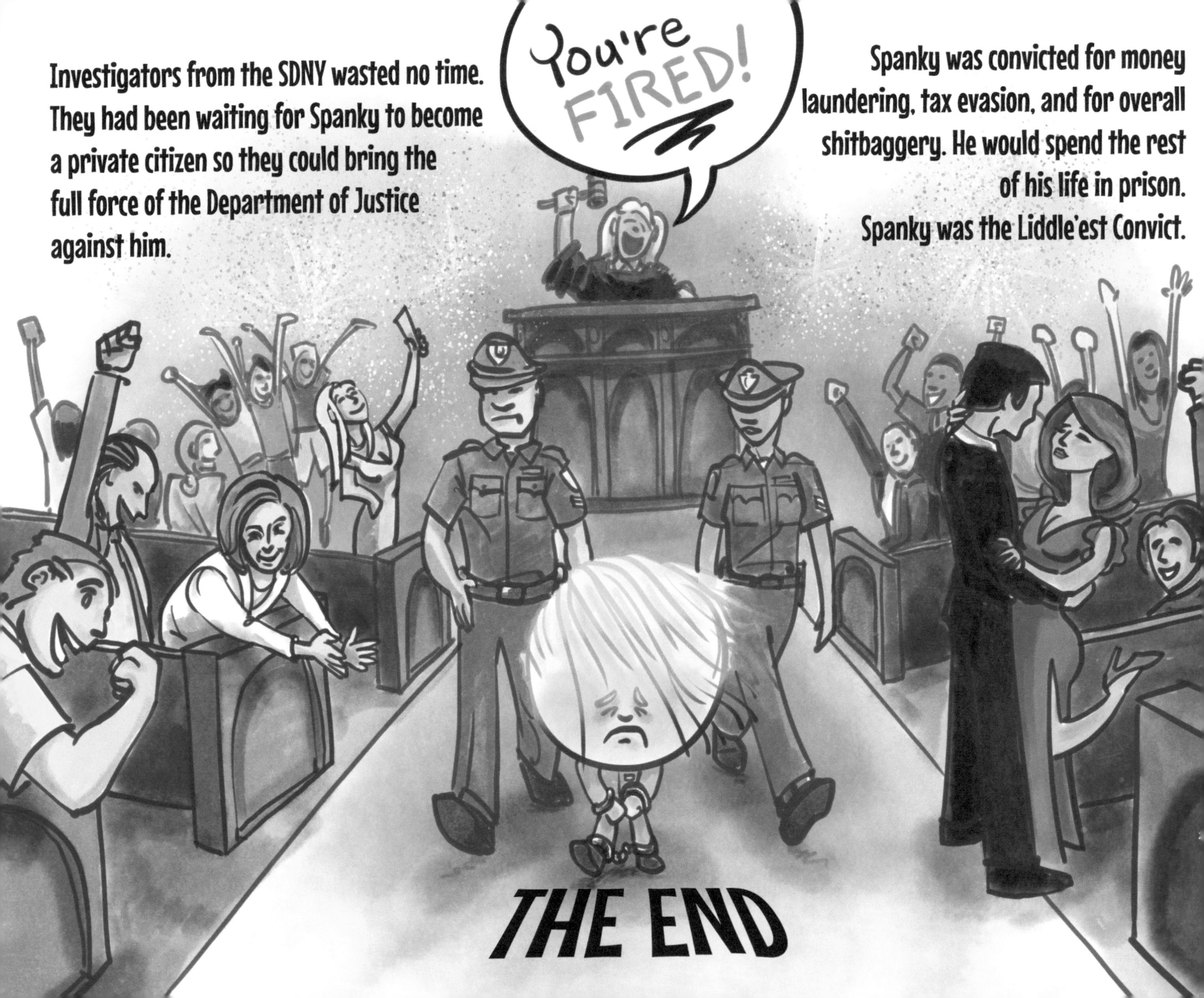
Investigators from the SDNY wasted no time. They had been waiting for Spanky to become a private citizen so they could bring the full force of the Department of Justice against him.
You're FIRED!
Spanky was convicted for money laundering, tax evasion, and for overall shitbaggery. He would spend the rest of his life in prison.
Spanky was the Liddle'est Convict.
THE END

MORAL OF THE STORY

Don't be a lying, racist, disgusting, descpicable, traitorous, useful idiot, fascist demagogue, creepy dad who's a traitorous, Russian asset, Putin Puppet, and bad at business, TV, marriage, tweeting, and presidenting.

The lies, extortion, money laundering, and overall criminality will eventually catch up to you.

BELIEVE ME.

Printed in the USA
CPSIA information can be obtained
at www.ICGtesting.com
LVHW061434131024
793698LV00036B/242

Grade 4

two octaves ♩ = 72

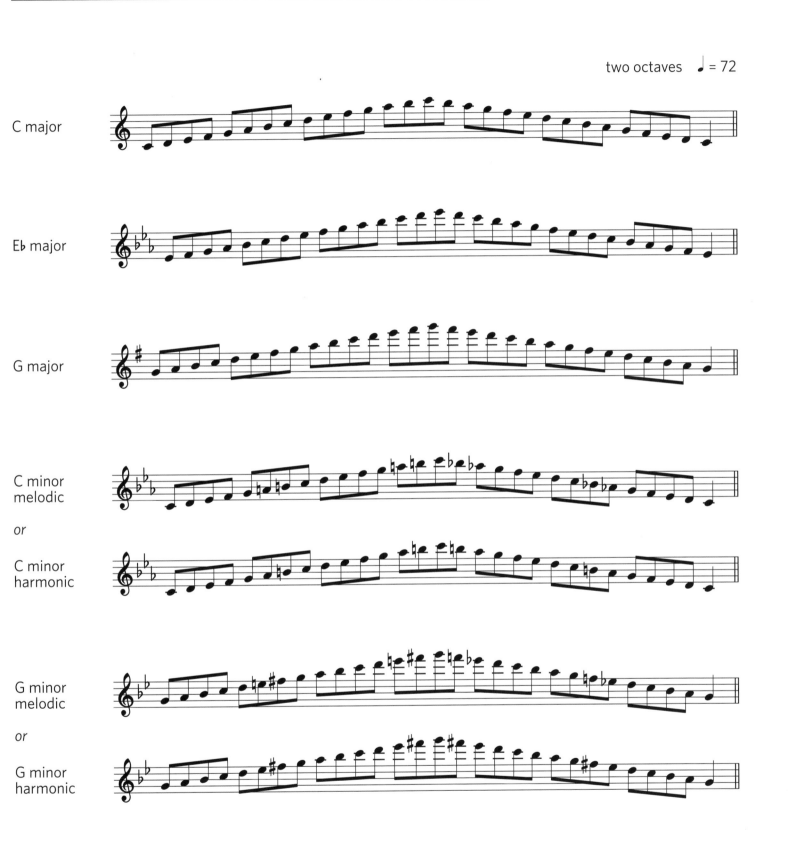

C major

E♭ major

G major

C minor melodic

or

C minor harmonic

G minor melodic

or

G minor harmonic

Grade 4

ARPEGGIOS

from memory
tongued *and* slurred

to a twelfth ♪ = 108

A major

B♭ major

A minor

B minor

two octaves ♪ = 108

C major

E♭ major

G major

C minor

G minor

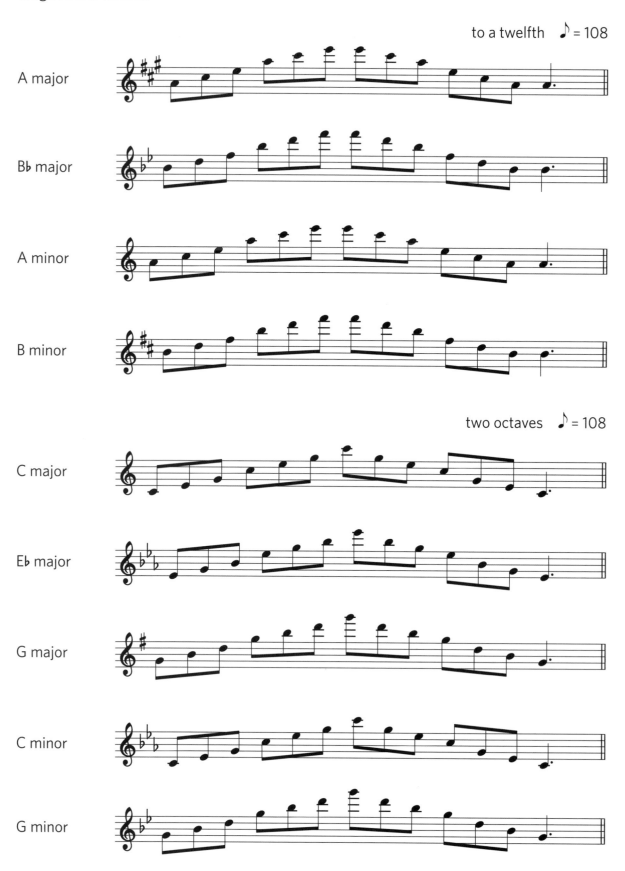

DOMINANT SEVENTH

from memory
resolving on the tonic
tongued *and* slurred

two octaves   ♩ = 54

in the
key of G

CHROMATIC SCALE

from memory
tongued *and* slurred

two octaves   ♩ = 72

starting
on D

Grade 5

SCALES

from memory
tongued *and* slurred

two octaves   ♩ = 84

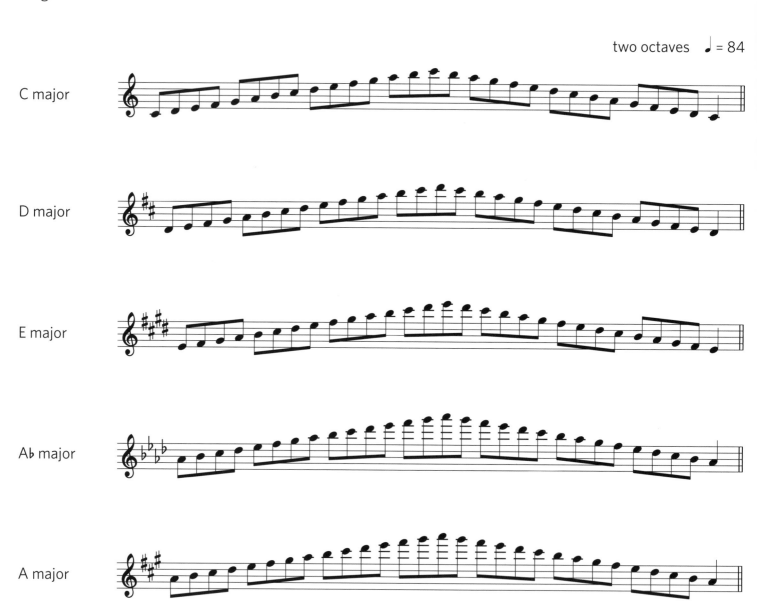

C major

D major

E major

A♭ major

A major

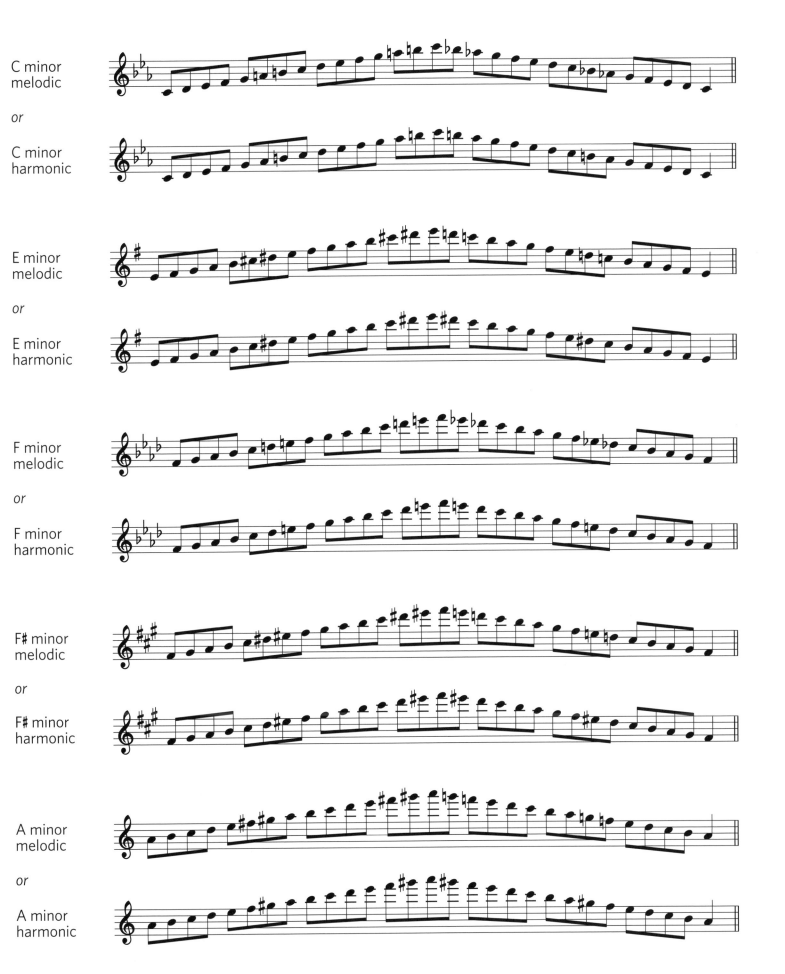

C minor
melodic

or

C minor
harmonic

E minor
melodic

or

E minor
harmonic

F minor
melodic

or

F minor
harmonic

F# minor
melodic

or

F# minor
harmonic

A minor
melodic

or

A minor
harmonic

Grade 2

Tempo di minuetto

21
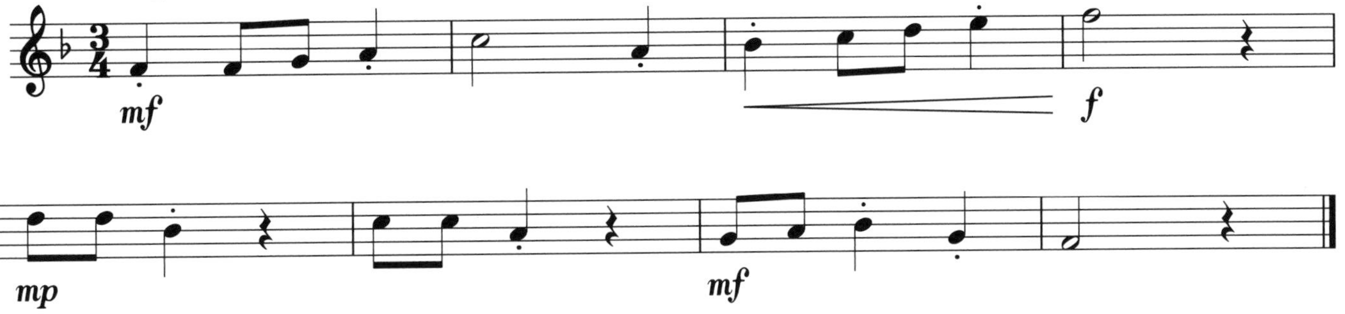
mf
f
mp
mf

Moderato

22
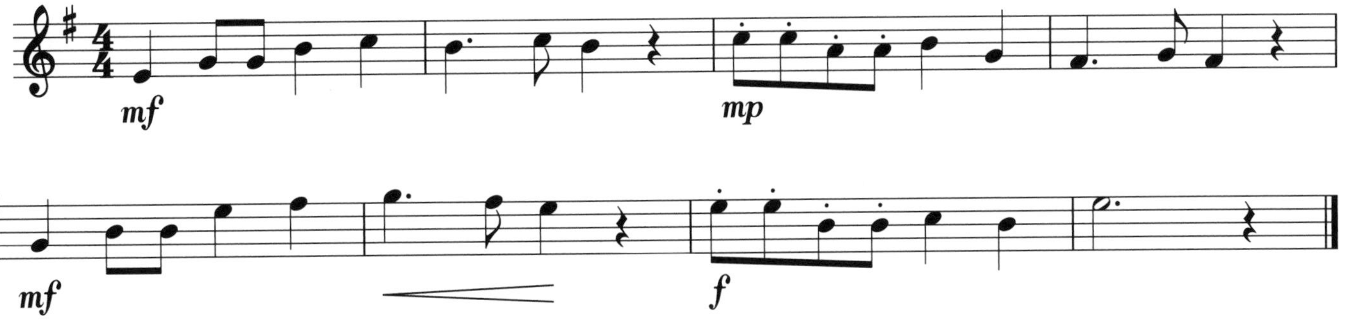
mf
mp
mf
f

Allegro

23
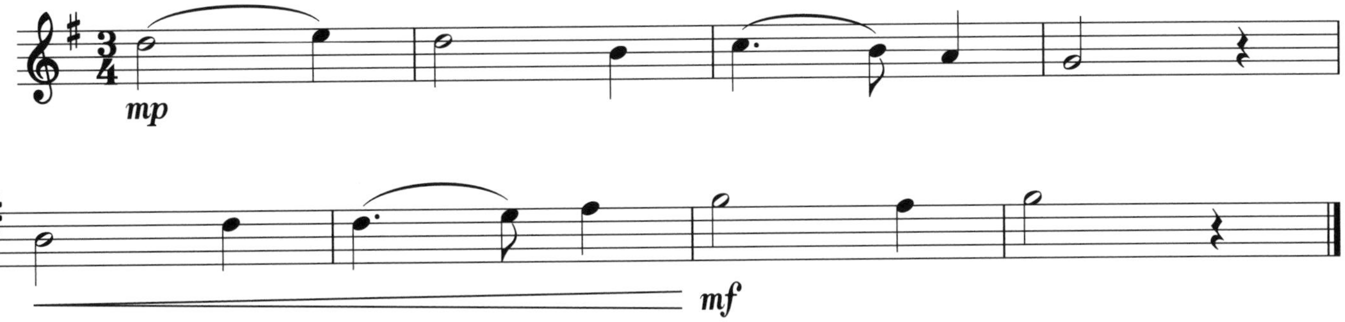
mp
mf

Dolce

24
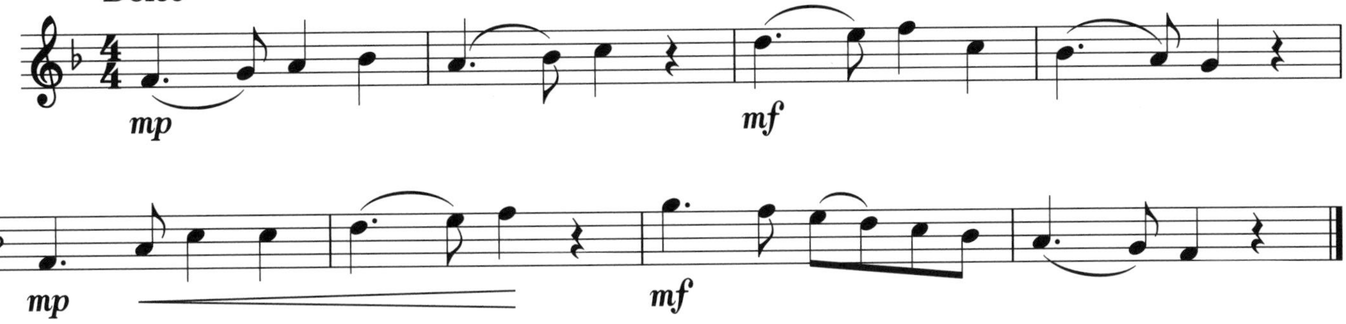
mp
mf
mp
mf

Grade 3 SIGHT-READING

Valse

1
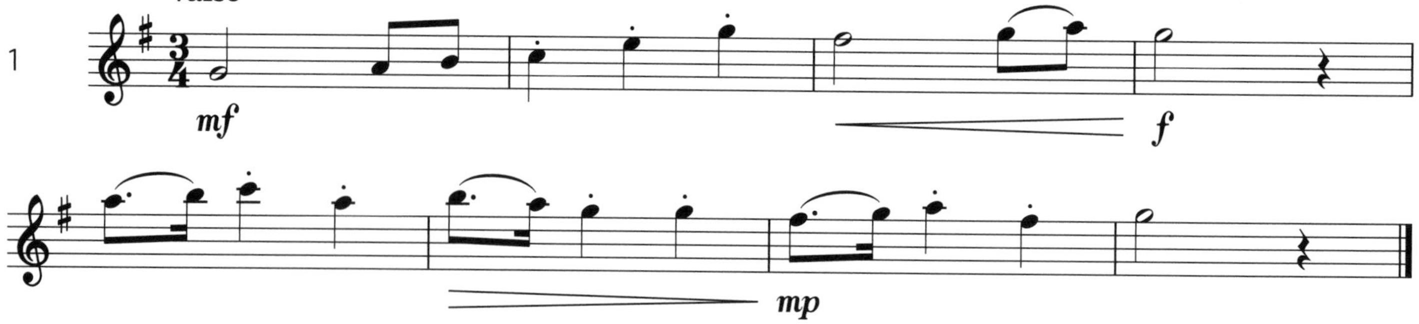
mf
f
mp

Risoluto

2
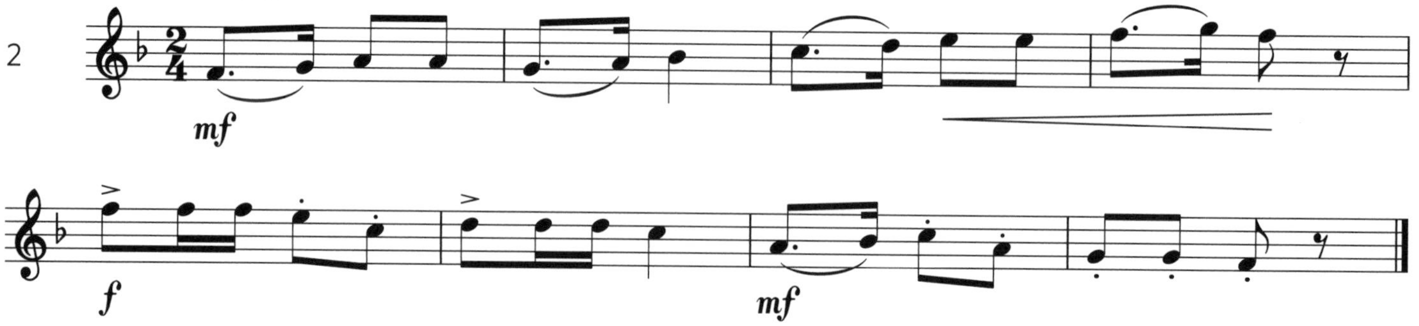
mf
f
mf

Playfully

3
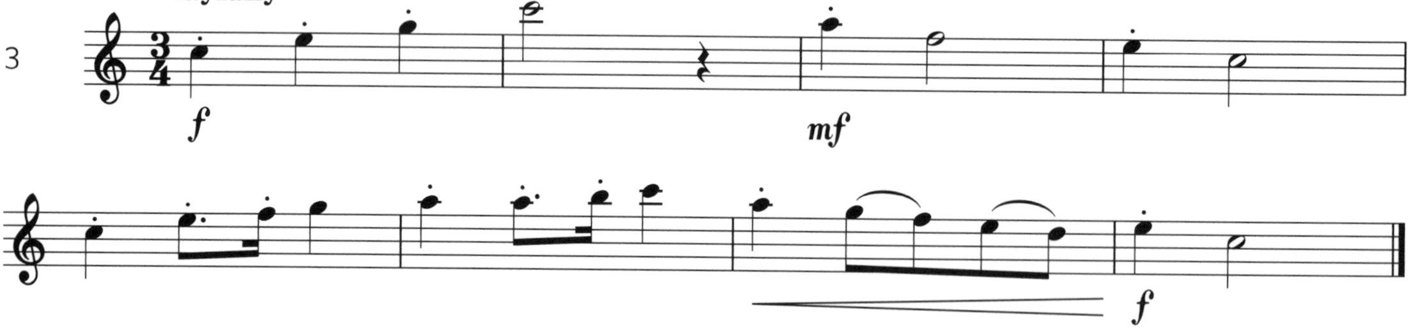
f
mf
f

Espressivo

4
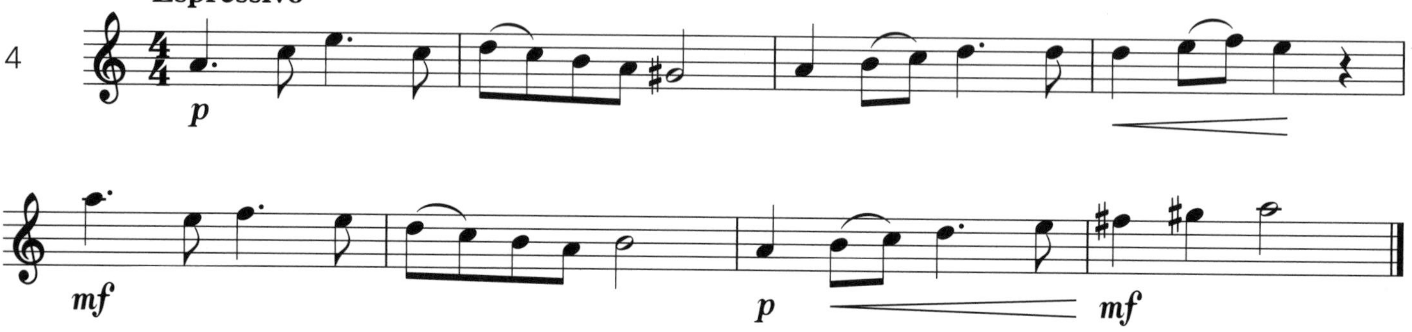
p
mf
p
mf

Grade 3

Andante

5

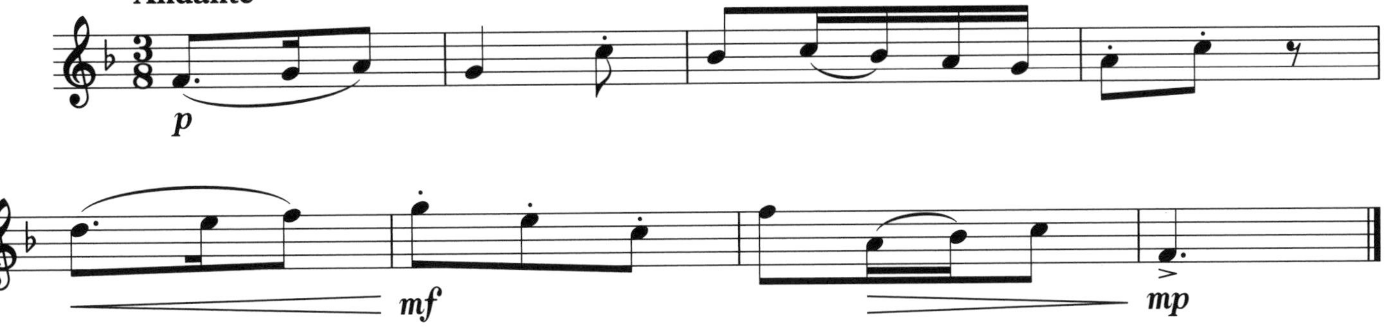
p
mf
mp

Grandioso

6

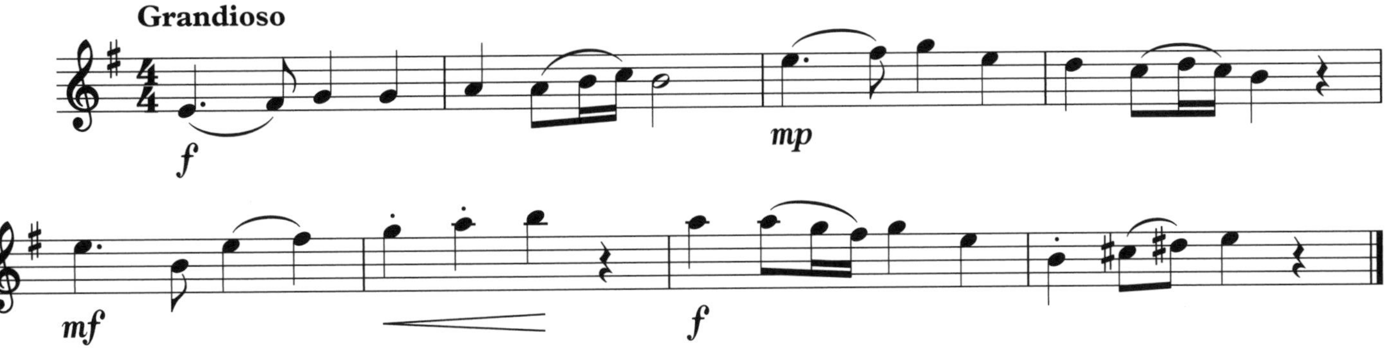
f
mp
mf
f

Energico

7

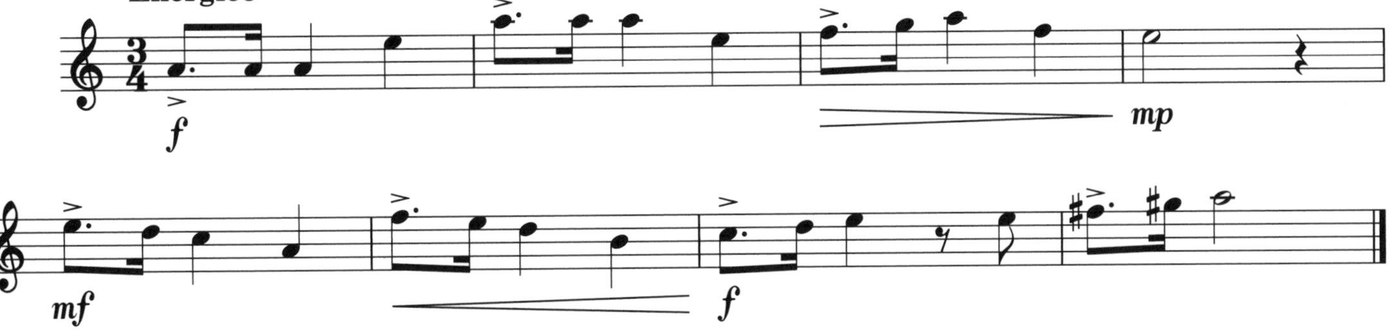
f
mp
mf
f

Alla marcia

8

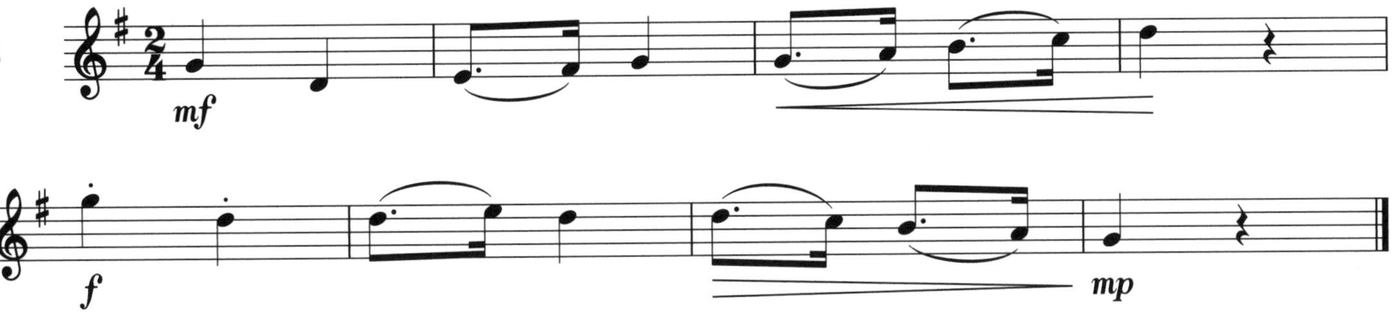
mf
f
mp

Grade 3

9

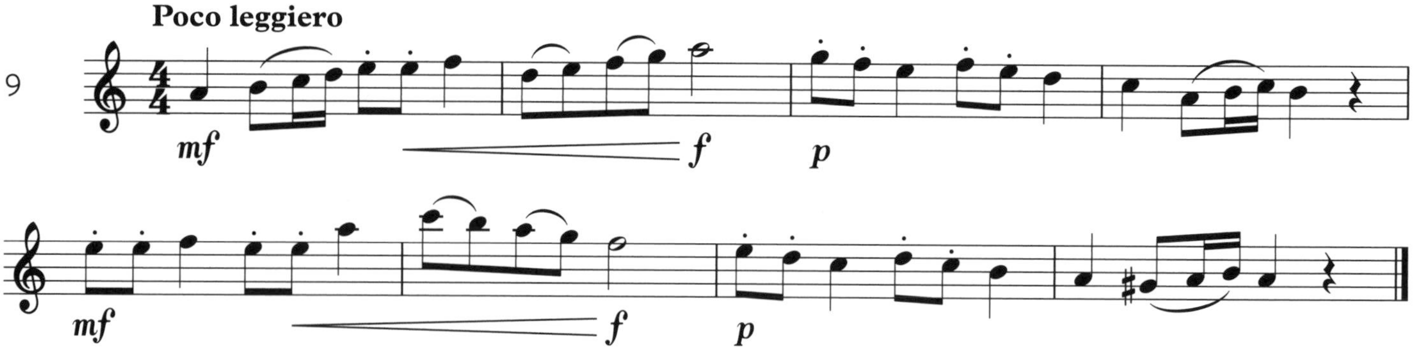
Poco leggiero
mf
f
p
mf
f
p

10

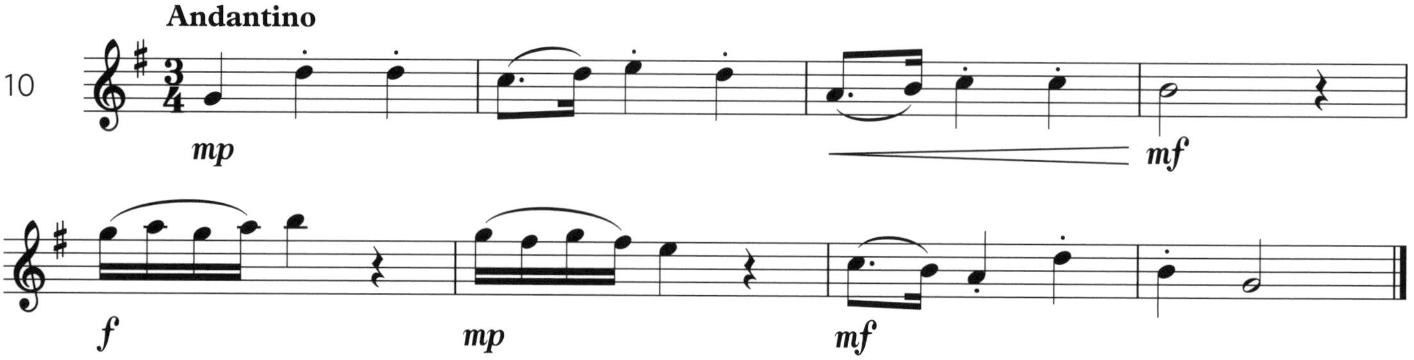
Andantino
mp
mf
f
mp
mf

11

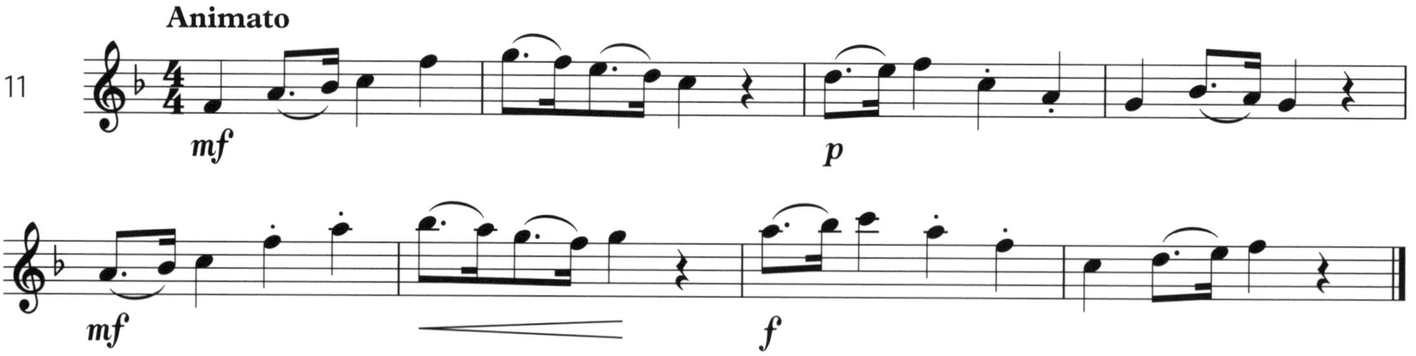
Animato
mf
p
mf
f

12

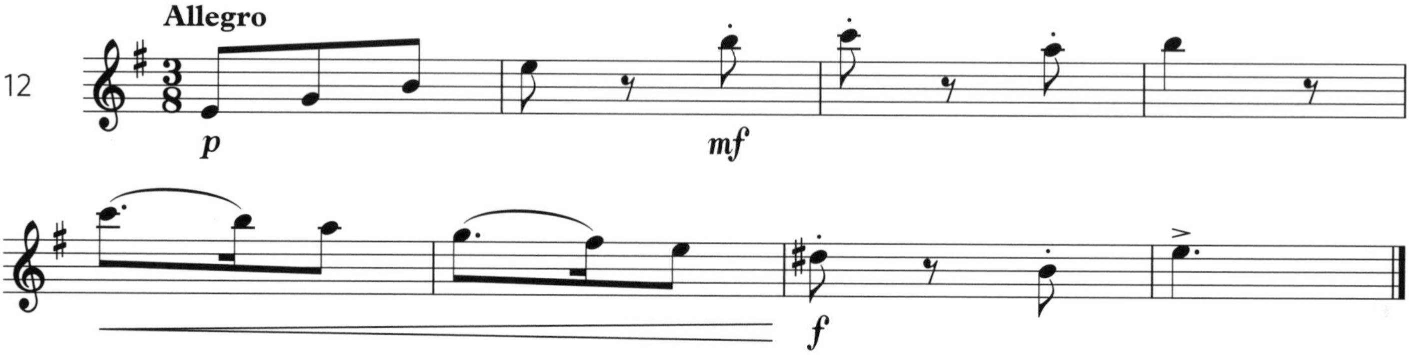
Allegro
p
mf
f

Grade 3

Waltz

13

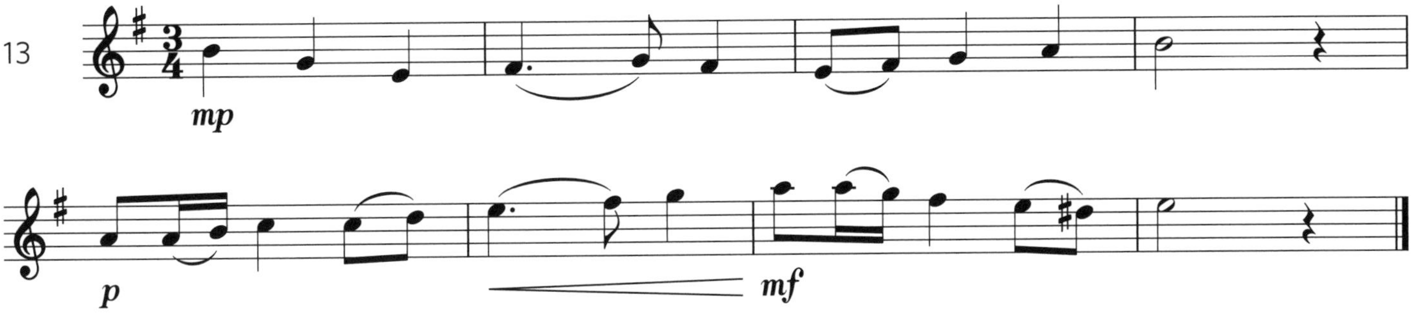
mp
p
mf

Giocoso

14

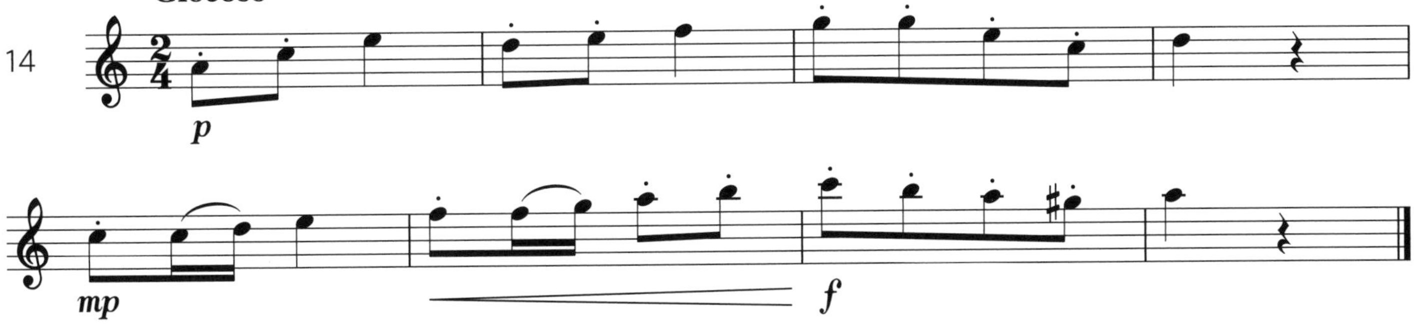
p
mp
f

Grazioso

15

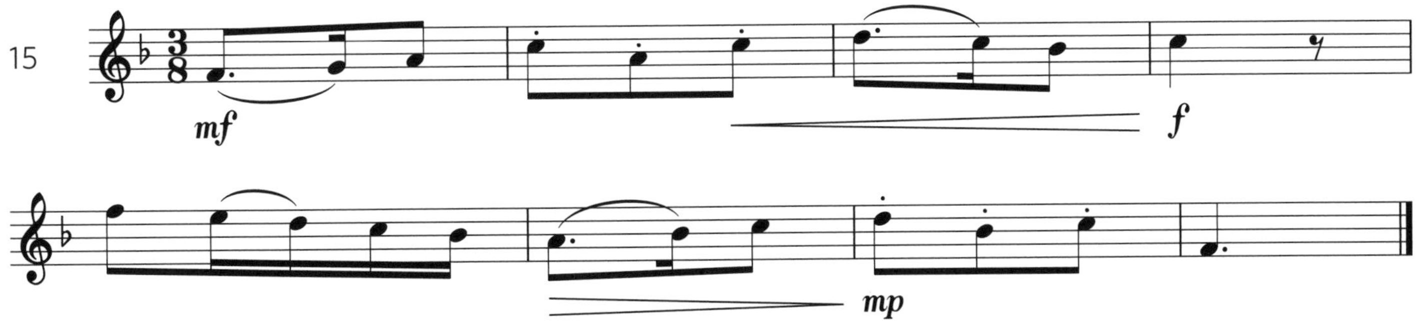
mf
f
mp

Risoluto

16

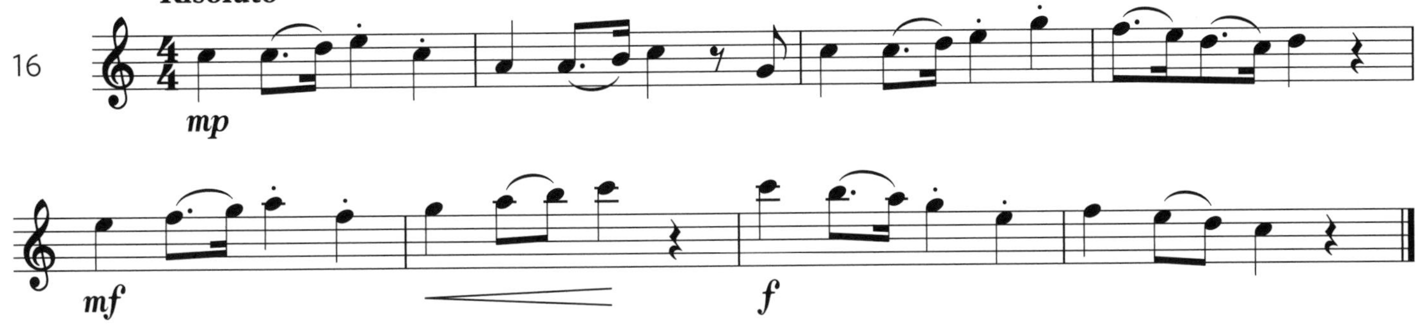
mp
mf
f

Grade 3

Allegretto

17

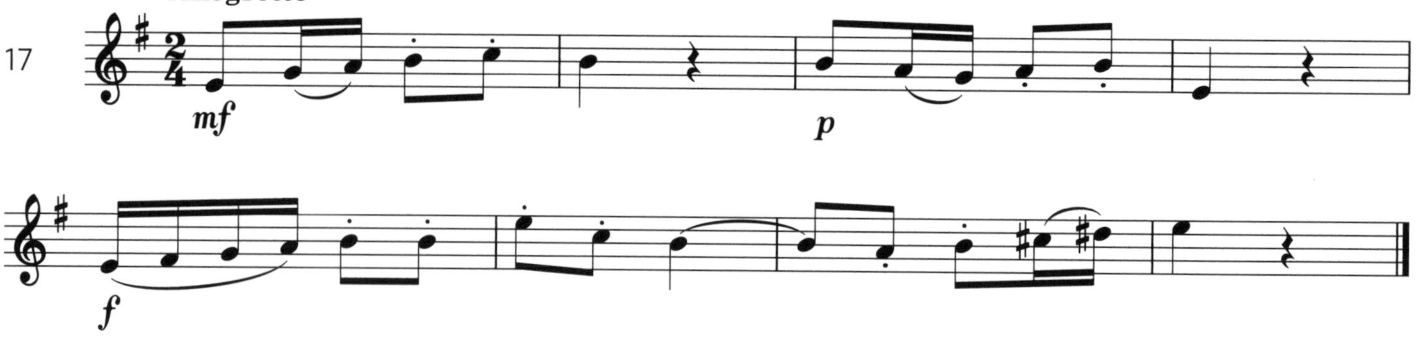
mf
p
f

Grazioso

18

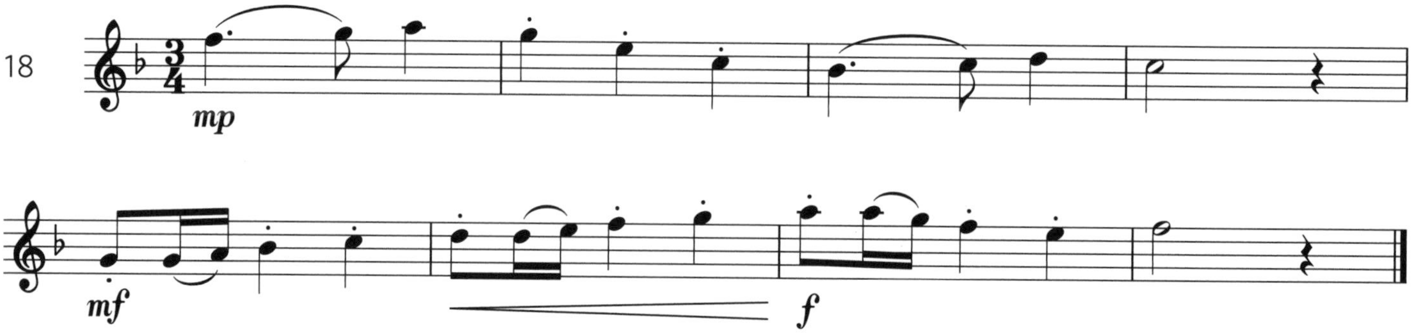
mp
mf
f

Alla marcia

19

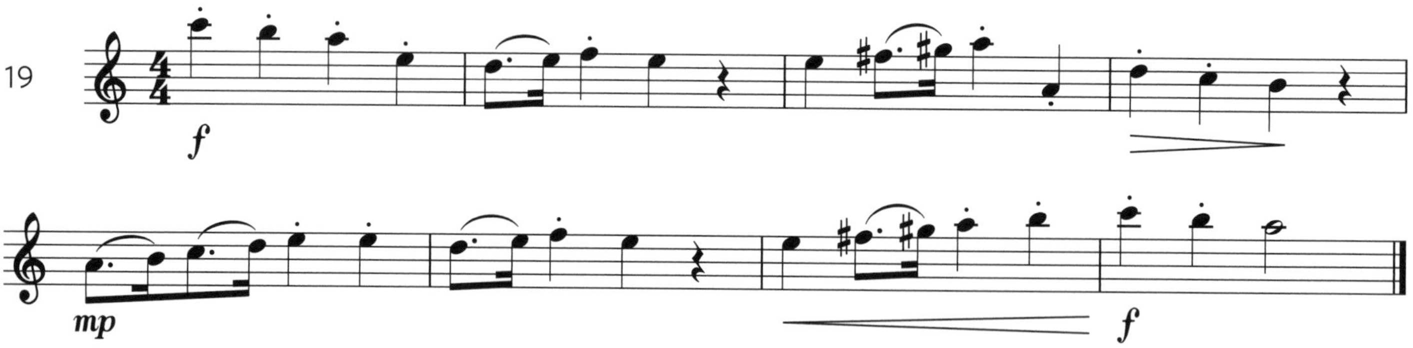
f
mp
f

Animato

20

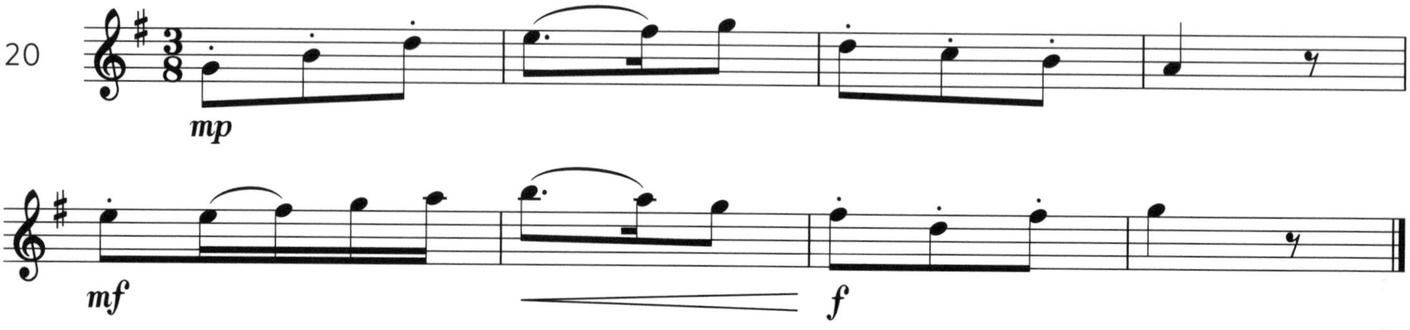
mp
mf
f

Grade 3

Espressivo

21

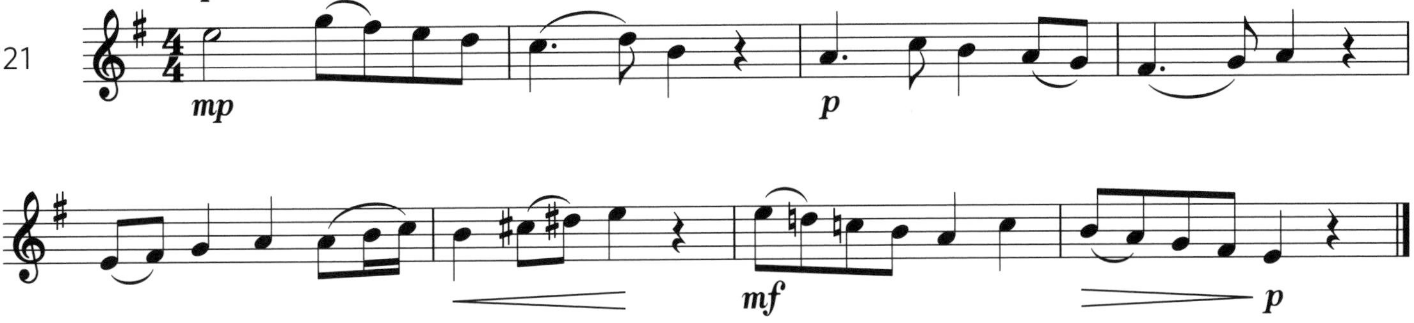
mp
p
mf
p

Energico

22

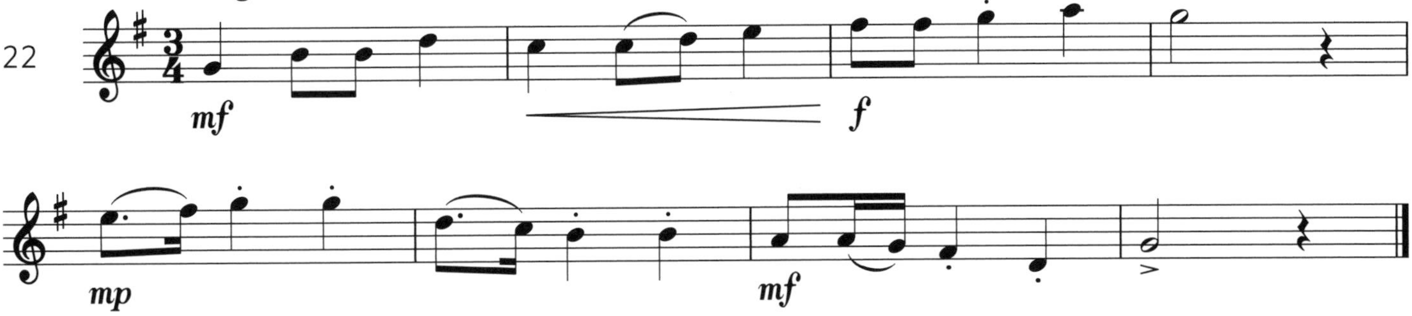
mf
f
mp
mf

Gracefully

23

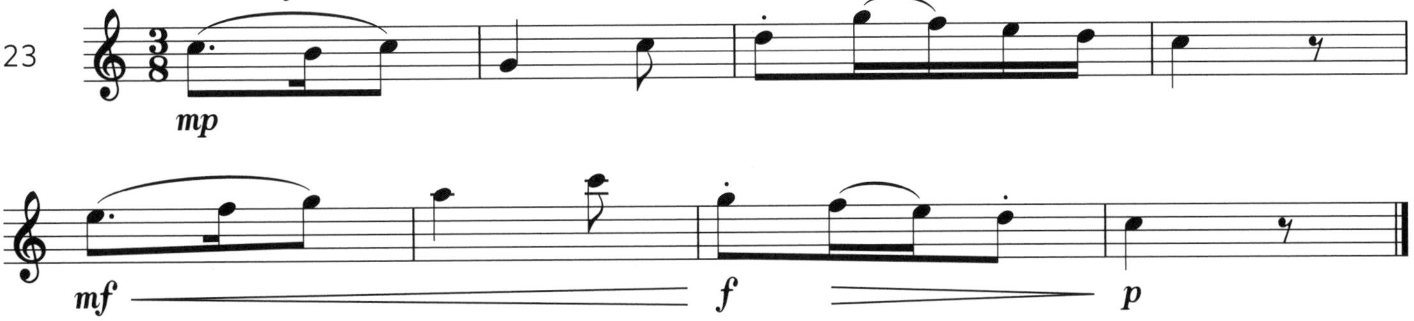
mp
mf
f
p

Andantino

24

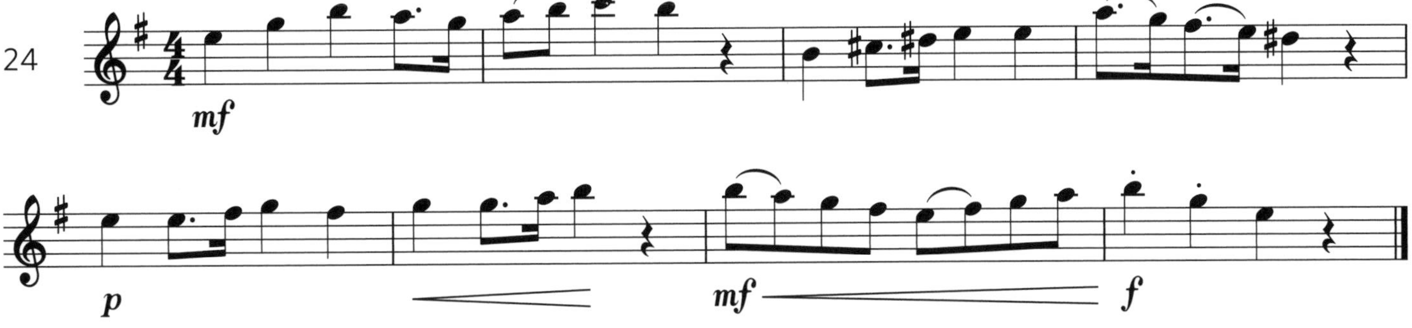
mf
p
mf
f